Bio-Energy Diagnostics

Bio-Energy Diagnostics

◆

Methods, Procedures, Techniques

Elena Bakalova

iUniverse, Inc.
New York   Bloomington

Bio-Energy Diagnostics
Methods, Procedures, Techniques

Copyright © 2001, 2008 by [Elena Bakalova]

All rights reserved. No part of this book may be used or reproduced by any means, graphic, electronic, or mechanical, including photocopying, recording, taping or by any information storage retrieval system without the written permission of the publisher except in the case of brief quotations embodied in critical articles and reviews.

iUniverse books may be ordered through booksellers or by contacting:

iUniverse
1663 Liberty Drive
Bloomington, IN 47403
www.iuniverse.com
1-800-Authors (1-800-288-4677)

Because of the dynamic nature of the Internet, any Web addresses or links contained in this book may have changed since publication and may no longer be valid. The views expressed in this work are solely those of the author and do not necessarily reflect the views of the publisher, and the publisher hereby disclaims any responsibility for them.

ISBN: 978-1-4401-0115-1 (pbk)
ISBN: 978-1-4401-0116-8 (ebk)

Printed in the United States of America

Contents

Bio-Energy Diagnostics i

Introduction XI

Preface 1

CHAPTER ONE

The Human Aura 3

CHAPTER TWO

To Bioenergy Diagnostics 27

Description Of The Methods for Diagnostics

CHAPTER THREE

Bioenergy Diagnostics Of The Emotional Body 37

CHAPTER FOUR

Bioenergy Diagnostics Of The Electric Counterpart 43

CHAPTER FIVE

Diagnostics Through Consecutive Remote Bioenergy Palpation. 51

CHAPTER SIX

Implementation Of The Method Per Systems 61

Digestive system 68

Endocrine system 83

Excretory system 91

Genital system 95

Mesaraic system and locomotorium 100

Nervous system 106

Diagnostics of sensory organsThe Eye 120

The Ear 125

The Nose 128

Cardio vascular system 129

Spleen and lymphatic organs 136

CHAPTER SEVEN

Diagnostics Through Remote Bioenergy Palpation Of The Fourteen Basic Energy Channels 139

CHAPTER EIGHT

Diagnostics Of The Chakras 165

CHAPTER NINE

Use Of The Visualisation In Diagnostics 175

CHAPTER TEN

Techniques For Organising And Balancing Of The Energy 181

CHAPTER ELEVEN

Technique For The Projection Of Light And Increasing The Energy Field 187

***

Elena Bakalova is an able and well-read healer. She has achieved undisputed results in the diagnostics and treatment of plexitis, hernia discalis, and arthritis, of some lung, gastro-intestinal, kidney and gynaecological and other diseases. She normalizes the blood pressure and brings to balance the energy in the human organism. Elena possesses the extra sensory capacity to see the spectrum of the aura of the human body.
Assistant professor Vera Tocheva, in "The extrasensory perceptive persons in Bulgaria", Military Publishing Complex "St. George the Conqueror", 1992

***

Elena is not merely an extrasensory perceptive person, but an interesting distinctive psychologist, who helped me to understand new facts about the boys. Elena is helping any and all of them to achieve in a desired moment a state of self-defence, inviolable space, in which they can accomplish the instant of concentration. If in 1960 someone waved his hands around us with serious intentions, we would have fainted with laughter. Now I think that the bio-energy healers are part of the future of sports.
Nourair Nourikyan, double Olympic champion in weight lifting, coach of the Bulgarian team of weight-lifting, in "The extrasensory perceptive persons in Bulgaria" Military publishing house "St. George the Conqueror", 1992

***

With your gift and talent I am sure, that you help many patients. I am ready to learn and apply the proposed by you methods for diagnostics in my medical practice. Most humbly I invite you to work together in Thailand.
Dr. Thabatchai Krisanaprakinkot – Thailand, Thanksgiving letter for a report made by Elena Bakalova at a congress in India in February 1996

***

Mrs. Elena Bakalova applies alternative methods of treatment with very good results in cases of malfunctioning of the mesaraic system and the locomotorium, metabolic and endocrine diseases. She possesses insight and understanding for the patient and every single manipulation of hers in the process of treatment is filled with care and kind attention.
As a charismatic person, she possesses extra sensory capacity to see the dynamic changes at an early stage, as well as to use the bio-energy in combination with other alternative methods as a therapy.
Mrs. Bakalova possesses good knowledge, experience and education and this makes her a valuable assistant in the medical practice and research.

Dr. Savvas Kassotis - surgeon, Athens, 1998

***

"Mrs. Elena Bakalova has helped me with diverse alternative therapeutic methods to overcome a very serious paralysis of the legs and pelvis, resulting from a serious trauma.
I think that the application of alternative medicine in combination with medical therapy shortens the time for rehabilitation and has better results. Mrs. Bakalova possesses vast knowledge; she is well trained, and is able to help to a great extent the medical work."

Dr. Ekaterina Kirinauku, Anaesthetist, Athens, 1998

***

From the first moment I met Elena Bakalova, I felt an inner power and radiation that possessed me completely. I was intrigued with how easily she could diagnose only with her mind, whilst we, doctors, subject our patients through a series of examinations and still many times find it hard to draw any conclusions. I was stunned when she was capable to measure a patient's pressure from a distance, which was confirmed right after the measurement, and she also knew how to improve the patient's condition of health.
Beginning from the knowledge that our mind can either cure or harm the body, Elena Bakalova takes us in an unforgettable journey, in her book, discovering the function of our inner self and showing us how to unlock the secrets of our healing capabilities.
Through a series of imaginative thoughts, meditations, confirmations and self-healing exercises, we can explore our self and find a personal

route for the balance between the body and the mind, which is the key to a healthy life.

Elena Bakalova reveals that the more you understand yourself, the more capable you will be to develop your own diagnostic and therapeutic powers.

Thank you, Elena.

Dr. Anny Sotyropoulos
General practitioner- Gastroenterolog

***

Nothing is more real than real stories. So let me tell you a real story. I had a problem with my kidney. To stick to the truth – I did not know that something was happening to my kidney, all I knew was I was going through terrible pain. I had to do a medical test. And then another test, and another. It took two weeks to understand that I have a stone in my kidney. I was informed about the size and the place of the stone. It took two weeks, but at least /at last/ I knew.

Then I met Elena. I have a problem…, I said. "Yes, I see – she said and pointed her hand. – You have a stone in your right kidney." Then she told me the size and the place of the stone. It took her two minutes. No pain, no lost time, no ruined nerves, no X-rays. We opened the file with my medical tests and compared the results – they were identical.

I don't believe that someone is seeing with someone's hands – I said – please, tell me the secret. "No need to tell you, she said, it's all in my book, you may read it. And you may do it yourself. If it is a secret, I want to share it with everybody. That is what the book was written for."

I read the book. I did not start healing other people, obviously I am not that caring and committed type of person like Elena. But I learnt a lot from her book. I learnt that I was not right to divide myself into body, mind and soul and wonder which is of paramount importance. Elena Bakalova's book taught me all is equally important and more – I am in charge of myself, my body is my right and my responsibility. This didn't change other people's life, but it changed mine. What more can a book do?

Vesselina Sedlarska, journalist

x

Introduction

Studying, 25 years ago, the module of Pathological Anatomy in the University of Palermo in Italy, I discovered a big secret that no professor taught from his chair: "the human body has the innate ability of self-healing. When the balance is disturbed, illness comes to light". The shock is great. For what reason was I studying this kind of medicine that did not take into consideration this big truth?

Reading all the scattered pieces of Hippocrates medical methods that are saved, I found out that the truth was known then. Through the centuries, it got forgotten to be discovered again from some "pioneer" researchers of the 20th century.

Graduating, I needed money to continue with the speciality of Gynaecology specialising in the field of extracorpular insemination. The symptoms of life brought in my way an Italian that had just been anointed a bio-energetic healer (an unknown capacity to me then) from the college of Zannata. In order to get a government licence to practice, he needed a doctor to supervise the course of the patients. I accepted, more out of curiosity for research rather than for financial purposes. During one year of our collaboration (I then quitted because I did not have the time to be up for so many shifts at the hospital) I wrote down cases of cured patients that my medicine classifies as psychosomatic.

Nothing peculiar up to this point, considering that the patients believed in him. One incident, though, created gaps and questions (that remained unanswered for many years) about my knowledge until then "Scoliosis of a high degree was completely restored in 7 sessions – confirmed from x-ray examinations before, during and after".

Years went by applying as best as possible for me all that I had been taught to couples and women that sought my help. In 1995 I suffered from total loss of 2 of my 5 senses: smell and taste.

With two babies, my every day life was very hard. The diagnostic examinations led me to a dead end. There was no apparent, noticeable or countable pathology. The colleagues advised cortisone treatments for a year, without any guarantee about the outcome.

It was then that I seriously considered an alternative approach for the restoration of my senses. I recalled my experience in Italy and started looking for someone in Greece that knew the use of bio-energy. A friend of mine cautiously suggested Mrs. Elena Bakalova. At my first visit she diagnosed lack of Silica – a fact confirmed from a colleague doctor that utilises the salts Schoussler. After 10 sessions smell and taste were completely repaired. That is how I found out the capabilities of Mrs. Bakalova.

That experience was the stimulus for me (a real kick) to broaden my horizons with everything relevant to medicine. Now I know: Medicine, the way it is taught in the university, is not panacea and those who practise it are not gods. There are other kinds with a different method of approach that can contribute in the restoration of balance (therefore of health) via their routes.

Hippocrates' medicine (sought from us by those who originate from India! And are let down with what they find here), homeopathy, ayuverda, acupuncture, reflexology, bio-energy, flower remedies Bach, Schoussler salts, hypnotism-induced retrospection, to mention some kinds.

In conclusion, for classic medicine to survive, people who serve it are called upon to widen their field of knowledge. The so-called HOLISTIC approach of man wins more and more ground in people's consciousness.

What Mrs Bakalova offers with her knowledge and her unique way, certainly makes today's leaders of medicine in Greece smile. What we do not know or understand is not ridiculous nor should be scorned. I plead and challenge the practitioners and the researchers of medicine –those with doubts- to look upon the phenomenon and study it.

There are, indisputably, gains from the combination of knowledge from all different kinds of approach towards the understanding of the perfect machine of the human body.

There is of course a small percentage of researchers whose horizons are wide open and study the phenomena and the results derived from the use of bio-energy. For them I am positive that this report about Mrs.Bakalova's knowledge constitutes a real treasure. After all, there are enough centres already on this planet, where the masters of classic medicine use the results deriving from the study of bio-energy and fully co-operate with bio-energy healers with amazing effects in various fields- not only in health.

May this is a wish and vision for a better future for the Man with higher goal!

Dr. Ioanna Pavlou
Gynaecologist surgeon and sexologist

Preface

I have been writing this book for years. Every day: in my mind, in diaries, in notebooks, on pads. It grew slowly, changed its character, style, even its title. What didn't change was the address of my message - and now, the way it was long ago, I sincerely wish that it should reach to those readers, who never get tired to seek and discover new sources of knowledge and skill.

I dedicate this book to my patients, who endow me with trust, love and appreciation and always have helped me and have participated in my search.

I dedicate it to all doctors and healers, who were and are next to me, to all friends, who trust me and to whom I cannot explain in every language "how I do this".

I dedicate it to assistant professor Vera Tocheva, who is not among us any more, and who was the first to urge me to describe the diagnostic and therapeutic methods, which I practice. We began together to ponder on this book. I keep her notes on the first project of the manuscript as a dear reminder, charged with a lot of faith, enthusiasm and constructive energy.

"Bio-energy diagnostics" is my gift to the WORLD FEDERATION OF HEALING domiciled in London, which acknowledged my merits and accepted me as their member.

My idea is this book to pave the way for the one to follow, in which I would like to describe my practical treatment activities, about the application of some of the alternative methods of treatment with bio-energy. I will depict in it treatment procedures for specific diseases of the human organism.

I have come to the conclusion, that the time to hide and disguise in secrecy our knowledge and gifts has passed. To write books, without explaining how we heal and what exactly we do, so that to keep it secret. That is why I would like to narrate what I do, what and how I do it. In the book you will find very few cases from my diagnostic and therapeutic practice. I have included only instances, which illustrate the depicted methods and procedures.

The proposed diagnostics is absolutely harmless; it is based on the discoveries and achievements of the ancient and contemporary medicine and could be applied together with the classical methods of diagnostics. It is a road to self-cognition, opening the perspective for further mastering of the gifts set in every one of us. It reveals new possibilities for every person not only as an object, but as an active participant in the struggle for our own health.

This book is neither entertaining, nor light, nor easy to read. It is not varicoloured with curious happenings and paranormal phenomena, although I have been their object. Unique in it is the fact, that I, as one of the numerous bio-energy healers in Bulgaria and world wide and who have mastered my experience and abilities with labour and efforts, with the riches and inherited wisdom of ancient schools, I WOULD LIKE YOU TO KNOW HOW THIS COMES OUT.

This book is for the ones, who together with me will come to believe in their own ability to come to know and heal at least themselves, their families, especially their children. It is never too late for such an attempt: to recover your health and the harmony in your own body, in the magnificent and beautiful vessel, destined to be the habitat of your beautiful soul.

Are you ready for this road? Let us tread it together, armed with the most precise instruments, possessed by mankind from the creation of the world until now - our senses.

CHAPTER ONE

The Human Aura

The aura... This word for a very long time evoked in me the notion of an angel, surrounded by a coloured cloud. Or I linked it to the golden nimbus around the images of Christ and the saints in the church, as a sign of divinity and superior spirituality. And I did not have the slightest idea about the link, not to speak about the identity of the aura with the human body. Many years later, probably after 1980, this naive idea of mine started to change due to publications in the periodic. I worked in a library and had the opportunity to follow everything that came out in the field of space travels, parapsychology, Eastern philosophy, and bio-energy. As my interest was growing, I began to follow the scientific and informative publications. With time I discovered new for me terms such as ethereal level of matter, electro-magnetic field of the human body, animals and plants, extra sensitivity (ESP), psychotronics, bio-energy processes, reflexology, Chinese acupuncture system for treatment... I read everything I could lay my hands on, discussed what I read with my friends, and even dared experiment what I have learned from books. Those affected were, as you can guess, my next of kin.

Very important for my progress were the contacts and my meetings with people, engaged in non-traditional medicine, with psychotronics, with investigation of the extra sensory capacities of man. They gave me the opportunity to feel that I am not alone, that a fellowship of gifted with exceptional capacities individuals, willing to implement their gifts and at the same time seeking understanding, support and safeguard is being formed in Bulgaria.

This need of intercourse gave birth to the societies of psychotronics in almost all big cities. The support of the state was materialized in the establishment of the Laboratory of Bio-energy with the Ministry of Public Health, situated in Plovdiv.

It was the time of great expectations, of enthusiasm and hope that a way will be found for the recruitment of ESP with proven capacities in the system of the Bulgarian Public Health.

I felt the need to work more and more and to improve my skills. At any opportunity I took part in seminaries, causeries, training courses, congresses. After each of these events I felt more-knowing, more confident and more able.

As a turning point in revealing my extra sensory capacities I consider a seminar, organized by the Society of Psychotronics and the Bio-energy Laboratory in Plovdiv. At the seminar participated, among others, several people from Bourgas: Dr. Kasarov, the radio engineer Stanchev, the extra sensory perceptive person ESP Sonya Georgieva, chemical engineer by education, the journalists Katya and Anna. We came to know each other at the bus stop of the Spa near Stara Zagora, and a few minutes later I already had the feeling, that I have met familiar people, whom I have known for years. While we walked to the hotel, they started to discuss the colours of my aura. I was stunned, that all could see my aura. Even in the beginning I thought, that they were joking, but later I began to listen intentionally and each thought expressed by them automatically imprinted in my consciousness.

The days of the seminar turned to be unforgettable, even, I would say, a crucial for me experience, days for testing of my capacities and abilities. With the help of my new friends and most of all of the radio-esthesist Nicola Stanchev, whose faith in my capacities was so convincing, I achieved then several things, which gave a new start of my progress as an ESP and as a healer.

At first I gained confidence in myself, in the power of my mind and in my ability to direct it, to change its direction, depth and its vibrations, to perceive, memories and process huge amounts of information, to pay attention to everything and at same time to classify it, by separating the significant from the insignificant.

The second thing was the diagnostics through a photographic image. We conversed in the hotel room and suddenly Nicola Stanchev took Ann's passport, gave it to me and asked me to go out of the room, to make diagnostics only through the passport photo and, when I am ready, to come back. It was for the first time I heard about such a thing, I asked him to clarify what am I to do, what he was demanding from me, but he refused by assuring me that I could do it alone. I was not sure, but I felt awkward to refuse. I stepped out of the room and tried to concentrate. I have made diagnostics from life, but the feeling now was very different. I pointed my right hand to the picture, starting from the head and as the picture reached as far as the breasts, I went on in my thoughts down along the imaginary body of Ann, observing what vibrations cause the organs on my middle finger. I had the feeling, that everything is normal, but I retried and I felt a painful pulsation in the gall-bladder. I entered the room and said, that Ann is healthy, but probably there is a small concernment in her gall-bladder and I asked to check this from nature. I still remember the pleasant feeling that came over me, when I found out, that my first diagnosis, made from a picture, was successful.

After that my new friends explained to me and showed me how to fix my eyes, in order to find the contour image of the energy counterpart of the human body. I started to exercise and quickly learned the techniques. In few minutes I was seeing it round the heads and bodies of everyone in the room. This outline was so ethereal and fine, as if the hand of an artist had drawn it with a sharply pointed pencil.

My strongest experience came on the following day. We talked with Katya in the hotel room. We were alone and all of a sudden I saw how around her body was emerging a beautiful golden light enfolding it from the outside and at the same time penetrating in it, lightening it from inside and making it transparent. The clavicles, the humurus, the ribs became visible. One clavicle was with some dislocation and slight distortion. I didn't dare to move, this picture came and went like waves from my eyes to her, but did not disappear. I could look at it as much as I wanted and with every wave to penetrate deeper and to see new details. Bewildered, I stared at the lungs, the heart, the other internal organs; I saw their colours and pulsations. I did not know how long this went on, but suddenly I felt terrific fatigue and I was frightened. We had to converse for a long time with Katya and the others, until I calmed down, in order to analyze the miracle that had happened: I have succeeded to scan a human body and see a human aura.

Left alone in the room, I again and again went through what had happened. How could all this happen, without conscientiously provoking it? I have asked myself this question, and in my memory came out reminiscence from my childhood, a mysterious image that used to scare me. It would happen in the winter. We had a cold room, a closet, where my family stored fruits, jellies, bottled fruits. After supper my mother often used to send me bring something sweet. I would open the door of this room and on the wall facing me would appear an image of a woman in golden light. I would scream loudly and run away, my parents would come, turn on the light, take what they needed and award me with the inevitable "chicken-heart". They never came to believe me, that on the wall of this room I could see an image of a woman. The explanation of this phenomenon from my childhood now lit me suddenly: Probably I have been seeing the luminous image of my mother, which I have projected with my own eyes on the wall of the cold room.

Months after the seminar, by order of the management of the woollen textile combine in Sliven, together with the radioaesthesist Nicola Stanchev, we carried out a survey of the labor conditions in the workshops of the Combine. With the methods of radioesthesis we made numerous measurements on the energy at the working place and the factors, influencing it, as well as its effect on the bodily and mental health of the personnel. We measured the rate of noise, vibrations, composition of air in the working premises and worked out proposals for the measures that would contribute for reduction of fatigue and for the preservation of the health and labor efficiency of the workers. We even expressed ideas for additional special processing of the fabric, which remained unrealized. But this is entirely another story.

In this period I discovered myself the radioesthesis as a science, learned to discern the vibrations of water, gold, copper, silver and other chemical elements, to work with the ruler of Turen, observed luminous spectral lines of acetone, chlorine, sulphuric acid. Not with special apparatus however, but with my own eyes. The illustrations of the spectral analysis in the chemistry textbook paled in comparison with the picture that I saw: forms and gradating hues of red, yellow, blue. I had the feeling that, day after day, I was discovering something new about myself. Together with another world - the world of light. I found even explanations to my previous "visions", which I was afraid to talk about even to my next of kin. Then I realized that the human bio-field is an extension of the physical body, that it has its own structure and parameters. The bio-field spans to a distance of 180 centimetres from the outlines of the physical body, but its size is variable, and from its dimensions one can draw conclusions about the health of a man and to reveal diseases, remaining hidden at routine medical research.

I remember that by this time I have managed to find the cause for the unexplainable fatigue of three women from the laboratory research of metals. While doing their job, as a result of their contact with metal samples, their bio-fields were shrinking by 30-40 centimetres each by the end of the working day. The same effect takes place as well with women, wearing too much gold, silver and other metal adornment for a long time.

I needed time, in order to realize, that our, human eyes, possess the characteristics of a photographic camera: they register the object and project it in colours, reciprocal to the real. How could this happen? Fix your eyes for a long time on a red rose; concentrate your attention on it. Close your eyes and you will see the same rose green. Move your look to the opposite white wall. There, the rose again is green, but with bigger dimensions. Move your look to your knees, covered with a white cloth. Here, the rose appears again green, but with smaller dimensions as big as one on a post card. You can do the same with a lighted candle. Upon lighting the candle your look is fixed on the lighted match. Your visual analyzer fixes on the light and wherever you look in the next moment, you will project the received light with reciprocal to the known to us light-spectrum colours. The actual colour of the flame of the lighted candle is yellow. If you close your eyes and look afterwards ahead, you will see it green or blue-green like the colour of the peacock plumes. The same comes out when you fix and concentrate your look on the man in front of you. You can see his luminous radiance and accordingly to blow it up or reduce it. The realization of the dependency between the distance to the fixed objects and the projection of their luminous image has helped me to determine the safe distance for work with every single patient.

Now, years later I am convinced that the capacity to feel the electromagnetic vibrations of the bio-field and to see the colours of the human aura, can not only be mastered, but also be perfected. This capacity is set in all people, but very few find the audacity to free themselves from the interdictions and restrictions, which they have imposed on themselves and to find the key to the new knowledge.

The teacher Peter Dunov in his research on races, which had populated and will populate the Earth, writes, that with the coming of the epoch of the Aquarius conditions and spiritual prerequisites for the creation of a sixth human race will be created. During the time of its development, a new sense shall be formed in man - a special mesh, through which he will be able to perceive the fairy world of space. There is an unknown energy of light, which is outside our sensory range. It would not be possible to encompass the new ideas that are to come through this culture without the help of this light. He says: "When this mesh is formed, the human sensorium will perceive the second octave of the colours. There are organic and mental vibrations of the colours. For the perception of the new light we must be ready. The old light is only an antechamber to the new light".

According to me the aura sight could be developed, of course, to a different estate.

The first stage is the capacity to discern the object from its energy, to see the energy of the mountain, of the trees, of the whole surrounding nature. This is achieved through systematic exercises, on which I will not pose here.

The second stage is control of the opportunity at any time, when we want to, to see the energy of air, of the space in front of us. This is a magnificent picture. In it, vibrating, shine and twinkle thousands of sparkling spots in various configurations. Then we could find out how many and different hues of white there are in the astral world.

The third stage of the aura sight control persons, who can conscientiously fix and project the luminous radiation from the human body, from live organisms or natural objects. They see the whole spectrum of the human aura and can most generally interpret the information it contains.

The fourth stage is the controlled and deliberate use of the aura sight for achievement of research or medical purposes.

As I already pointed out, the aura sight is cultivated in many people. However, many of them are afraid for different reasons to admit this capacity of theirs. Researchers have found out that the babies discern their mothers by the light they radiate. Some ten years ago, the seven-year old boy of an acquaintance of mine asserted that he sees coloured lights on the back of his grandmother. His parents accepted his words with incredulity. But when I talked for a long time with him, I found out that the boy had determined very precisely the painful places on his grandmother's body. According to him at these places he saw a red light.

The parents of a seventeen years old girl had her confined in a neurological clinic, because they didn't believe her admissions, that she sees colours round the objects and persons and had decided that there is some kind of a mental disease. I had hours on end to convince them, that there is nothing unnatural in the gift of their daughter. In the past year the girl graduated and already works as an assistant in the higher institute. She keeps on improving her aura sight and uses it in her practice

In another case the colleagues of a woman paediatrician regarded her as mentally ill and their attitude forced her to undergo treatment for a long time in a psychiatric ward, because with her aura sight she found out immediately the problems of the ill children. Even now she keeps on utilizing this capacity of hers, but will hardly ever dare again to share this with her fellow doctors.

I have met people, who only for minutes under my guidance start to see their own electric counterpart. I know others, who see it, but dare not admit it and prefer not to talk about this. Third, after a prolonged period of therapy, observing healthy dietary regime and reading suitable literature, all in a sudden reach the cherished enlightenment. I have met even people who say: "Just help me see the aura and I will make you rich". I am not inclined to believe them because they expect someone for their sake to make the effort, necessary for the achievement of this goal. The waking up of these capacities cannot be bought with money, but needs a long work for spiritual perfection. Only then could be achieved the sensitivity necessary for the new sight.

I feel with my hands and see with my eyes all this and according to the state of the bio-field and the colours of the aura I draw my conclusions for the bodily, mental, psychic and emotional state of the man in front of me. And it is not only me. This ability is possessed by more and more people and their number will grow.

When I see the human body in a cloud of light with the seven colours of the rainbow, I remember one quotation from the Bible, in which God addresses the people in the Noah's ark, after the deluge, with the words:

> "I place the rainbow in the cloud,
> to form a sign of eternal testament
> between me and Earth".
> Genesis, First book of Mosses, Ch. 9:13

> "And My rainbow will be in the cloud
> and I will see and will remember
> the eternal legacy between God and Earth
> and between every soul, living in every flesh,
> existing on Earth".
> Genesis, First book of Mosses, Ch. 9:16

Every flesh, existing on Earth is a soul, blessed and shielded by God. Like the Universe is an aggregate of what is objective, unreal, natural and spiritual, physical and metaphysical, so man is a unity of spirit and matter. But the leading, essential, the eternal, the divine in man is his spirit. Accepting this idea, the contemporary scientific thought gets the opportunity to go even further in its progress and to give new answers to questions that have agitated mankind for centuries on end.

According to Jean Prior "man in his essence is a spirit. Let us not say, that he is a physical being, possessing spirit, but that he is a spiritual being, possessing physical body. The spirit - this is the man himself, the real man, and not an essence without form, a ghost without substance. "

"Man is triune; he is a spirit, spiritual body and physical body. The spiritual body is the vehicle and instrument of immortality. It is the real body, the true form of man, the one that does not change.

While the physical body is renewed from year to year, changes with age, grows old, vanishes, the metaphysical body does not change. It is the constant base of our identity, of our stability, of our personality."

Similar thesis develops the philosopher Annie Bezant. In her book "The Man and his bodies. The Universal Religion" she writes about the necessary emancipation from the fallacy in general currency, identifying the sensorium, which is our "ego" with our bodies, which are only its temporary vehicles. "As a man I understand the living, conscientious, thinking ego, the individual; as bodies - the diverse shells, with which the ego is clad, thus any clothing gives an opportunity to the ego to function in a separate field of the Universe."

According to the Bulgarian researcher Koubrat Tomov, the living beings are structured by two components: physical (atoms, molecules, cells, tissues, organs) and field (elementary and sub-elementary particles and heir respective fields). And the bio-field is composed of two interpenetrating parts: with a known physical nature and with an unknown nature. Namely the bio-field is the direct vehicle of the psyche and the sensorium, the information about man and about everything, what happens with him is coded in it. "It comes out of the limits of the body, theoretically disseminates to infinity and interacts with the respective physical fields and bio-fields in the whole space. Through their bio-field people interact among themselves incessantly and form the uniform bio-field of mankind - people are more closely linked, than through their external interactions. "

Before I go on with the components of the human aura, I find it necessary once more to point out my belief in the inseparability of man from his bio-field or from his aura. When I think or speak about the aura, I have in mind not something, which man possesses, but the man himself. The aura is the unification of his bodies, a form - a part of man.

I feel, see, and realize man, (the human aura, and the human bio-field) as an aggregation of material and ethereal elements, functioning as a system. In this system are included:
- a) dense (physical) body;
- b) ethereal body or the electric counterpart of the physical body;
- c) astral (emotional) body;
- d) intellectual (mental) body;
- e) causal body and
- f) spiritual body.

Each of these elements of the human aura possesses different density, frequency of vibrations and color. Mastering our capacity to intercept signals of the aura, to analyze precisely the received information, we will learn more about the processes taking place in the human body, we shall be able to form for ourselves more complete and more precise idea about the health, the psyche, the sensorium, the spirituality of every man. And to help him if he wants to and if this comes within our capacities.

I will try briefly to make a description of the elements of the human aura, guided first of all by my practical experience and my observations.

THE PHYSICAL BODY is the biological system securing the functioning of the human organism and its adaptation to the terrestrial conditions of existence. It is a dense, tangible, substantial whole, which is situated in space and possesses color, symmetry, limits, and weight. In it are situated, arranged and function in harmony the self supporting organs and systems. Every organ and function of the organism has its spectrum of emission, bearing specific information about its work.

The physical body is a biological system, arranged and organized according to the laws of space and time. The renewal of this system is made through biophysical, biochemical and physiological processes. The changes in it are a result as well of the practical activity of every individual in the physical world or field. As an organic system, it is sensitive even to cosmic radiation, food, heat, water and air.

In time and space the physical body functions inseparably of its electric ethereal counterpart or ethereal body.

THE ETHEREAL COUNTERPART is a repetition of the outlines of the visible physical body in space to the utmost details. According to some scholars the physical body and its ethereal counterpart form the so called energy level of the human bio-field.

In the ethereal body are reflected all life processes of the organism, it has electric and magnetic field, different stages of light and thermal radiation. Its direct energy link with the physical body is maintained through the seven energy centers, about which I will speak in another place.

The physical (dense) body and its electric counterpart are mutually bound up and inseparable. The character and quality of this link change in synchrony, according to the state of the organs and systems of the physical body. During disease of the physical body the vibrations and the color of the electric counterpart are changed. Here I would like to point out also something exceedingly important from diagnostic view point: before the beginning of the disease of one or another organ of the physical body, in the electric counterpart appear signals, which warn about the oncoming disease. This, I would determine it as anticipating the disease effect, gives an opportunity for prognostic diagnostics.

If we observe the electric counterpart in the day, we shall notice how it penetrates and comes out of the outlines of the physical body and forms a very fine outline up to five centimeters from it. When moving our look along the limits of this outline, we receive information not only about the present state of the organism, but also about previous or eventual future diseases. Every interruption of the graphite line is a signal of a problem - an old trauma, a fracture of the bones of the limbs, a missing organ. If one of the limbs for example is missing, its electric counterpart is present, however without the light contour line. Just like the described in medicine "phantom pain" which feel people with a cut off limb at the place of the missing leg or arm.

We could determine air as one of the fuels for the human organism. The correct breathing is very important for health. If a man has mastered the secrets of wholesome breathing, he controls one of the most important factors for preservation of his health. This is best shown in his electric counterpart - by the strength of his vibrations and brightness of his luminous radiation. In such a case the sensation for breathing remains at the background, the man is captivated by the feeling of calmness, security and unity with everything around. In this state, his breath or "pranatha", as the yogis call it, flows down the nerves of the body and feeds them, releases them from the tension and helps them to sufficiently realize their functions. At such moments the ethereal substance, which flows over the body and it's, drawn as if with a pencil limit could be best observed.

In her book "The Conspiracy of Aquarius", Marylyn Fergusson writes, that "with the course of time our bodies become our itinerant curricula vitae, which betray to friends, as well as to the unknown peoples the small and the big stresses of our life". This very successful metaphor applies mostly to our ethereal counterpart. According to my observations, it carries the whole information also about the diseases, which we have inherited from our ancestors.

The state of the electric counterpart is in direct dependence also with the food which we eat. I have observed people, who have tried to clean their organism with light and vegetarian food or have switched to fruit and tea reduction regime of nutrition. Already after the third day the color of their electric counterpart changes and becomes orange yellow, which is a convincing sign of the initial cleansing of the organism. Together with this occur changes also in the metabolism, relief from tension, fear and stress, accumulated in the muscles and the whole body.

The link between the physical body and sensorium on energy level is carried out by the ASTRAL BODY. It is known as well as the emotional body. It appears after the ethereal (electric) counterpart, but does not follow its lineament and form. Its stability in space, or its substance, is several times less dense than the ether of the body field. It could stabilize, increase or decrease itself, change its form and intensity depending on the energy of the feelings, which prevail. For example, in an outburst of rage it could acquire an indented form and be structured like a sparkling nimbus around the head. But with unanimity, concord and love in the family the astral bodies of man and wife fuse together, they have the same color and intensity of vibrations and so form the family aura.

Through the emitted signals of their emotional body, the teachers communicate with their disciples, and the lecturers and the preachers contact with their audience and nail their attention.

The astral body possesses one more, very important characteristic. It is capable to separate from the physical body for a certain time, to overcome significant distances and to come back. This transfer in space is made during sleep, meditation, during surgical intervention or temporary state of clinical death.

The astral body reacts and registers every thought, mood, wish, striving, by changing its characteristic. It percepts the positive or negative energy of the feelings and thoughts through the energy center of the solar plexus. Naturally occultists share the opinion, that the cardinal organ of the astral body is the big sympathetic nerve with its three centers or knots (plexus cervica, plexus cordiaque, plexus solaire). That is why, probably, quite often with very sensitive and emotional individuals a disturbance of the nerve supply of the organs in the abdominal cavity is observed. The pneumo-gastric nerve acquires coarse-grained structure. In this region they feel constant anxiety, "like a ball", as they say. The anxiety becomes their concomitant feature and they cannot calm down, in order to solve sensibly their problems. This "ball" becomes an emotional heart, which causes injuries successively of the heart, the pancreas, the gall, the stomach, the liver.

THE EMOTIONAL BODY appears in children after their third year. It develops under the influence of the emotional surroundings, in which they grow. It is formed, cleared and stabilized after puberty. As a mirror of the psychic and the emotional processes, it grows to the stage of organization in harmony with the cultivation of personality.

THE MENTAL BODY is a vehicle of the energy of the human sensorium. It works on and through the astral and the physical body. It is also called mental body and its energy - mental energy. In the physical aspect it occurs through the energy centers of the fontanel and the pituitary gland. I observe its stability in space or its substance as an extraordinary fine, oval, bright atmosphere (matter) in the region of the mentioned herein above energy centers. It does not follow the outlines of the physical body but has an egg shaped form open upward.

One could say that the substance of the mental body is so much as the brain has been cultivated. If we look at one comparatively uncultivated in mental respect man, it would be very difficult to discern its mental energy. But in individuals who have worked for a long time on their brain, have cultivated its capabilities, their intellect we find out that their mental body is sufficiently clearly outlined, in spite of the fine matter, from which it is made of, almost invisible for the internal senses of man. As a direct vehicle of the human personality, it grows proportionally to the mental development, becomes more and more defined and organized, shining in marvellous colours and vibrating with very high velocity.

The energy of the mental body depends also on the present state of the man and his brain. So for example, through the vibrations which I apprehend with the fingers of my hands I can determine the stage of the brain waves and how much it corresponds to the activity, which the examined person is occupied with at present: whether he has concentrated on his work, whether he is distracted and restless and so on. This could also be seen. If we compare the mental energy with clear spring water in a transparent glass vessel, the more concentrated the mind is, the brighter blue and crystal is the water in the vessel. So and the limits of the mental body are more clearly outlined with a shining band. When the brain works effectively and productively, then the substance of the mental body is clarified and bright. And to the contrary, when there are disorders in the brain activity, when the mind is not loaded sufficiently, when the thoughts are undefined, vague, elementary, the water in the vessel starts to lose its clarity.

The mental body of individuals, occupied with scientific and research activity and exceedingly of those, who have dedicated themselves to mathematical and technical disciplines, to exact sciences and medicine, is as if closed in a strictly determined form. With creative personalities in the field of the arts and culture, it is open. Probably because such people more easily succumb to their internal impulses, moods and feelings. The shape of the mental body is open also during meditation.

The mental body could be best observed during a conversation. Then the activity of the brain waves is high and the trained eye sees the marvellous transformations of the mental energy. It is exceedingly interesting to watch from aside a discussion with two or more participants, as well as the mental body of a lecturer during a lecture with the auditory.

It happens to me quite often during a conversation with someone to ask him: "What did you just think about and did not tell to me?" The answer usually is: "How do you know?" Years ago I found out, in a way unexpectedly for myself, that I can catch the moment, when the thought is formed in the mind of my interlocutor. Its nascence is accompanied by a flash of a spark. A kind of "clinking" occurs in the mental body, as if someone is striking a match. If the man speaks immediately this thought, "the spark" is transferred in the region of the Centre of Brocca (centre of speech).

However, if the thought remains unspoken, its light is not lost; it remains as a trace in the mental body, while its energy seems to inhibit for a certain time the state of the man. Quite often in such cases he loses his thought or the so called "white spot" emerges.

In 1994 I read about a research carried out in the Institute of the Human Brain in Saint Petersburg, Russia under the leadership of the academician Natalie Behterev. The scientific team found out a luminescence in specific zones of the brain during mental work. A brain space of six millimetres was fixed and three spots in it: for reaction to the meaning of a given phrase, for its grammar and for generalization, that is for analysis and synthesis. Zones, corresponding to different types of mental activity were found out and even zones, reacting to mistakes.

The information I read about this discovery made me really happy. I felt satisfaction, because it confirmed my observation for the luminescence of certain zones of the brain at the time of its activity.

Sometimes stress situations, conditions connected with exorbitant exhaustion of nervous or psychic energy, tragic events which have occurred suddenly can be the cause for temporary or permanent disorganizing of the brain centers and even to lead to cessation of their normal activity. The brain connections are disordered, the system is blocked. In such cases normalization of the brain activity could be achieved with bio-energy impact. And for comparatively short time, at that. I want to point out one such example.

It was in 1992 in Athens, during the congress, dedicated to the Sylva Method. In the "Zapio" hall, at one of the afternoon meetings I demonstrated to the participants the techniques, which I use for the concentration of the athletes from our weight lifting national team. Then in the hall came in two women clad in black and a burly, about sixteen years old boy. The boy's mother asked for an apology from the participants and explained that they have come to us with the final resort for help. She told us, that her son, George, for the last six months couldn't talk, neither read, nor write. They have visited many doctors, psychiatrists, psychologists - but without result. The breakdown had happened, when a relative of theirs rang over the phone and left a message, that George's grandmother, who had been treated in a clinic is in a critical state and expected to die.

Until then I haven't worked in front of so many people. On one part something inside me made me touch the field of this young man, and on the other I knew that it is dangerous for me to work in a hall, where there are many unknown people with different attitude towards the alternative medicine. I looked in direction of assistant professor Vera Tocheva, leader of our group at the congress. She nodded in support. I paled a chair in the middle of the hall, placed the boy on it and from a distance of about one meter, started to seek for signals for the oscillations of the brain waves. The vibrations were weak, the colour picture - dark grey. I started with movements of my right hand to clear the right hemisphere of the brain; from there I passed to the top of the head, where the center of the written speech is situated. It was choked and inaccessible but I managed to unblock it and to lead outside the accumulated energy. When after this I passed to the center of Brocca, I was already sure that there will be a result. I took a sheet of paper and asked the boy to write down the numbers from one to ten. He did it. Then I asked him to repeat words and short sentences, which I pronounced. At the beginning with difficulty, but after that more and more confidently, the boy repeated correctly. The result was stunning - for the present in the hall, for myself, and most of all for George and his mother. Only in several minutes the boy started to write and talk again. George finished school, became an engineer in air conditioning. He started to write very beautifully. His mother used to bring to me his notebooks filled with texts written in the technical face and proudly praised her son's penmanship.

The vibrations of the mental body from the human aura correspond by character and intensity, as I have already pointed out, to the brain work one is occupied with. They are influenced by the state of the nervous system as well. If for example, a man is anxious or feels afraid, the vibrations are like strong pricks, and the luminous picture has an orange colour. A gradual cooling and shrinking of the range of the vibrations is observed as well. The individuals with gentle character, who are confident in themselves and in their capacities, diligently do their work and achieve without tension their aims, emit vibration waves, which slightly elevate the fingers of the ones who make the diagnostics. The luminous picture in them is from blue to mauve-purple.

The capacity of a man to apply efficiently his intellectual capacities, to overcome the challenges of his life and his internal controversies in order to live and be realized in the real world, depends on the maturity of his sensorium, on his wisdom. Even the ancient philosophers had foreseen the truth that the wisdom comes when man is able to see and judge the things such, as they actually are. But this can't happen quickly. Peter Dunov - the Teacher writes: "Every man is a specific fruit, which ripens at his time. No fruit could ripen in twenty of forty hours. Notwithstanding how much he wishes this he cannot trespass the laws of Nature. "

THE CAUSAL BODY is the vehicle of our maturity, of our wisdom. It is a disc that had saved the lessons, which we have learned in the past and on which we record our most important truths and discoveries, acquired in our present terrestrial life. The information with which we come and go...

It is the second mind body, inviolable store place of the human experience. All causes, guiding the conscientious activity of man through all stages of his development are coded in it. Everything valuable and permanent is situated in it, there are the embryos of all specifications of the human conscience, which are worthy for building material of the future and give the meaning of the human existence. The program of the human life from birth to the physical death is stored in it.

It is the energy, which overcomes the universal space and time and in the moment of conception comes to our parents. I determine it as "a delicate colourless membrane of thin matter". I see it as the energy of foetus in pregnant women still in the first days after conception and in children up to three years of age - as a shining, sunlight funnel, nozzle of which starts from the top of the head. In adults it could be is observed very seldom: in moments of relaxation, meditation, full concentration on some problem, in minutes of inspiration or revelation. Then behind the back of their head flares a disc of sun light big between ten and forty centimetres. This is only for a moment, but it is repeated several times, as the sunshine disc appears either behind the back of their heads or comes out in front of their eyes. In such moments to repeat the words of people, whose causal body I have seen, they have made some important for them discovery, have solved a very complicated problem; have found the answer, sought for tenaciously and with sustained efforts. But for me these are moments of compassion. If then I point my hand to the man in front of me I feel how a flame licks me and I discover by intuition the mechanism of the whole energy process: I understand that the brain is the instrument through which the energy of our thoughts is transformed in memory. Memory - securing the stability of our highest achievements through the causal body.

In the life of some individuals there are events, which cause rearrangement of the information whose vehicle is the causal body. Then it gets rid of the already unnecessary information, erases itself or codes new links. Everyone who has succeeded to come out of his body, has had the chance to realize sooner or later what this means. The same effect could be achieved through the techniques for cleaning of the memory developed by the spiritual teacher OSHO.

I want to relate an experience of mine, which gave me the opportunity to convince myself that such phenomenon "rearrangement of the information" exists. On December 9th 1992 in the evening, during full moon and full lunar eclipse, it happened so that I felt myself about one meter above my body. My hands felt it, and the feeling was as if it was not mine. I was not alone. My husband noticed my strange state and asked me: "How am I to help you?" His voice came from beneath, but I didn't answer. In spite of everything he started to rub my legs, the solar plexus, my fontanel, but in vain. I had the sensation that something strange is happening with me, inexperienced until now, but I was calm and sure, that nothing frightful will happen. Then I felt how for one moment my life went back through the years. I saw all the cardinal events and experience in my life until now to pass by in one line from the present moment till I was 20 years old. After this I was dominated by a state of lightness and inexpressible happiness, what I haven't felt till this moment. I was like one light blue sphere, which floats in the air with many other such spheres. I flew to another, greater and pink sphere, obsessed by a state of happiness and peace. On the next day in the morning at 8 hours and 10 minutes my energy returned in my physical body with a strong pain in the region of the left heel. The pain continued up my body and disappeared.

This case erased entirely my fear from death. I felt actually how the passage from one state to another takes place. I understood how the cleaning of the disc from the memory for all hard moments and bad recollections in my life has been realized. My life reeled with a great speed, like a movie, twenty five years back and even with details. I had the feeling, that a secret has been revealed to me "why do people remember nothing from their previous life and come again to the Earth purified". Because, before they die, they fall in the same state, in which I was this night. Their whole life reels like a movie to the moment of their birth and so erases the memory for it. In such a way the reel with the film of his life is torn and the man soars upwards. Like a rocket, which burns along its way its stages (bodies), in order to be able to soar higher and higher and convert into a star (energy) at a specific energy level. A man, when dying is converted in a rising star, and when he is born, this energy falls like a star. Up to this moment I thought that when a man is born a star is born.

Once more I convinced myself how good are the techniques of OSHO for retrospective cleaning of the stress suffered in difficult moments of our lives. I saw and realized the strength of the light in the world of the causal body. I felt the vibrations upon leaving and entering in the physical body and these up. I realized that there are moments in our life which do not depend on our mind and our will.

On the following day after what happened with me I discovered that I see with easiness the transient problems and foresee the future of unknown people...

In the energy of the causal body is coded everything significant, everything experienced throughout our lifetime. I have in mind here, on the Earth and in the other dimension. Everything which is essential and has entered the program of the causal body remains there forever. And it could be changed with great difficulty. It determines the character of a man here on Earth - according to the accumulated knowledge, skill and experience, according to the hardships and suffering, which one has experienced and which have helped him to find a meaning for his path.

There are moments of maturing and getting wise. When do they come? When we hear something about ourselves and it does not touch us. What do I mean? One of the most important conditions in order to feel the change is, when we hear an unjust assessment of us, a gossip, a slander not to let this hurt us. Simply, what we have heard, already does not touch us, does not cause a spasm and pain in the stomach, heart beating, sleepless nights, as it had happened years ago in such situations. This is because the criteria which guide our life are already different. When we lose something very precious for us, for example a precious family relic or priceless gift from a close person and it doesn't hurt us. We simply accept it: it had happened and we can change nothing. Nothing bothers us, life goes on. If we feel like that, this means, that the change had occurred and we are on the threshold of our spiritual perfection. We know who we are. We believe in ourselves. We know our capacities, follow our intuition and it leads us along the way, we have to pass...

In such cases appears the next body in our aura as well -THE SPIRITUAL BODY. This field corresponds to God – the origin.

Everyone, who is able and has learned to see, will notice it most often in newly born up to the 40th day and in individuals who have worked over themselves for a very long time and have succeeded to harmonize their life and their energy. The vision of this body always is connected with a tunnel of light upwards. I can call it and determine it also as a fountain, fireworks of light which passes through any physical obstacle and soars upward. Another thing which I see is the interchange of energy from below upwards and vice versa. A man at this moment sees one small speck at a great height on the top of this luminous tunnel and feels himself as a part of the divine, calm and happy. The energy in this tunnel in special cases could burst from above downwards and cover you with light. This according to me is a proof of the divine energy. Our aspiration to help ourselves and the others by organizing our biofield, is rewarded, our energy comes back by drowning us in a stream of golden light.

The spiritual body is seldom revealed. In my practice until now I have observed the spiritual body of a dozen of people. All of them have successful professional career. Among them there are doctors, musicians, actors, scientists. They have rich life experience and marvelous mentality. The kindness, selflessness and the nobleness are leading features of their life philosophy.

It is very difficult to depict the spiritual body, because the colors are different from these, which we perceive in our everyday life. It is a shining white light in different hues. The experiencing of this vision is so shuddering that sometimes I catch myself to utter thanksgiving words to the almighty for the gift, which I receive.

I will try to depict the moment in which I saw the spiritual body of a well known doctor. He is a surgeon and a perfect professional, often he has to operate seriously ill, when the life of his patients hangs by a thread. Alongside with this he has passed through the traditional schools of the holistic medicine, has mastered homeopathy and had helped me personally just for few hours to drop the high temperature. He is a master of the meditation and works daily for his mental protection with his personal models. In the end of each therapy I teach my patients to perform one very easy mental exercise for organizing of the energy. During such a case, as he had mastered to perfection the technique to concentrate himself and perform creatively the exercise, he transformed in one model. Totally synchronized in front of my eyes appeared all of his energy bodies. His physical body shined like a mirror, looked at from aside; the ethereal (electric) counterpart was like a fine opal glass, and their link with the astral body was maintained with one shining, perfect white lotus, coming out of the solar plexus. The lotus pulsated and from its center came out circles of white light, which widened and receded, filled the room with such clear and shining white light which I saw for the first time. I found out its vibrations, its fine, pressing in my fingers waves, which were as if helping me to step back and I had already receded to a meter behind his head. Looking upwards I found out, that the ceiling of the room had merged with the light. My eyes experienced this happiness to see and feel one powerful live embodiment of a man, who had taught himself to irradiate.

I told already that one to succeed to see the spiritual body is a rare phenomenon. In India however I saw how this is caused deliberately for different reasons. For example, in order to help the meditation or to achieve high level of concentration.

I will narrate one specific case during a congress with the participation of leaders in different professions from many countries all over the world. It was held in the university "Brahma Kumaris" in the mountain Mount Abou.

In the hall for common exercises in the morning and in the night gathered for meditation about 1000 people. Then it was filled with magic music and we meditated in absolute silence. In one of the evenings the participants in the congress probably tired from the tension during the day, were noisy and couldn't concentrate for the meditation. DADI JANKI, who had to lead the meditation, stood on the rostrum. The music sounded in the hall, while we kept on making noise. At this moment we couldn't concentrate at all and meditate. I looked to the left and to the right and although the light was very weak, prepared to write. My look was concentrated on the rostrum, where DADI JANKI stood. I felt that something extraordinary was going to happen, it was betrayed by her state. She was so frail and tiny actually but the light which she radiated could be compared to nothing. I intercepted her thought, I understood that she is taking a decision to prove that she could draw our attention and I started to watch her luminous radiation. She concentrated her energy in the cardiac chakra, there started to shine crystal clear amethyst light, it shaped itself like a sphere and started to depart from the body by gradually floating to the first row in the hall. All of a sudden all stopped making noise, deadly silence fell and my eyes stared above the first row. There the amethyst light vibrated and pulsated, transforming in concentric circles and they moved away and spread over the hall. Above the head of the teacher had risen a column of golden light, which obliterated the ceiling of the high hall. A real miracle, if you have the eyes to see it... I felt, that in the next moment the teacher will regain the energy of love and calmness, which she transferred to all present in the hall and then the amethyst sphere started to come back to her breasts. I foresaw even the moment, when she would cease the controlled emission of energy towards the hall and would rise from the armchair. So it happened. Later I shared my observation and convinced myself, that others from the spectators have become witnesses of the vision as well. DADI JANKI showed the force of love and calmness, which she mastered and has orientated to the spectators, in order to achieve a specific goal. The energy, which emitted this gentle old lady, had filled all the space and showed the strength of the great gurus, the miracles of spiritual energy.

I watched once the spiritual body of an orthodox priest in the church in one of the big hospitals in Athens, Greece. It was on the day before Easter. The previous night my daughter felt ill. We went to the hospital to make some tests and till we waited for the results, we entered the church. I got acquainted with the priest and asked him to read a prayer for the health of my daughter. After a conversation with her, he read the very prayer. While sending his prayer, a great funnel of golden light came over the priest. This light was pouring around his eyes and hands and drowned the body of my daughter, who had fallen on knees before him. The priest was as if transformed in one sphere of light about 3 meters in diameter. His voice soared and changed in a strange melody which pulsated with the vibrations of the luminous sphere. A feeling of calmness and peace drowned my soul because of this beautiful vision. I was happy of what I saw which revealed to me the boundless power of prayer.

When could we feel, that our spiritual body is being created? When in minutes of calmness or in the evening before falling asleep we start to feel how around our head a circle is formed, I could call it also a ring, of golden yellow light. This light pulsates, vibrates and heads upwards. While elevating, this circle diminishes its diameter and amalgamates in one spot. In this instant appears a second circle, which is also headed upwards. This is repeated many times but comes out neither very slowly, neither very quickly, so that its beauty could be observed and felt. Upon the elevation of these energy luminous rings upwards, a tunnel is formed above the man shining with a violet light. The sensation is that you are part of the Universe that you dedicate your vibrations and come in contact with it. There where the circles join at the end of the tunnel, stands a yellow luminous spot.

This moment could be reached also after long study of raja yoga with a dedicated teacher. The study of raja yoga helped me to peep in the astral world, to comprehend it, to accept it and master it as a reality, as one world, which penetrates and springs from the physical one. In this world there is no place for anxiety and fear. It exists and reveals to us thousands capacities for research and discoveries in different fields.

I tried to depict my perceptions, observations and opinions of the human aura, led first of all by my own practice. I omitted in most cases the opinion and experience of well known theosophers, healers, doctors, scientists in this field, who helped me to go ahead, who strengthened my belief in what I see and feel, and who gave me the opportunity to come to know myself and my abilities.

CHAPTER TWO

To Bioenergy Diagnostics

The contemporary medical science and practice has made great strides in the last few decades. The treatment of hereditary diseases and cancer, the transplantation of organs, the introduction of new drugs, the technical rearmament, the early diagnostics and prophylactics are achievements which no one could deny.

Still more and more discoveries of the molecular physics, biology, genetics, cybernetics, chemistry enter in the diagnostic and the therapeutic processes. With the help of the constantly renewed and sophisticated instruments and the computerization, surgical operations are already performed, which until lately were regarded impossible. The achievements of virology and immunology reveal more and more optimistic perspectives for mankind to get rid of the diseases, which accompany it for thousands of years.

And in spite of all this, or may be namely because of all this, more often voices are heard calling for the necessity of change in the approach, in the model of health care.

By attacking the pain and the disease medicine sometimes leaves in the background, neglects basic needs of its object - man. It is still less interested in the personality of the patient, in his spirituality, in the values which he holds.

Another big problem of the contemporary medical care is its price. It can't be afforded by the majority of citizens of our planet today. This fact makes inaccessible for millions in need the most effective methods and means for struggle with their diseases. The hope for real, effective and commonly accessible public health care still remains a dream for the people from the poor nations and states.

The necessary change must make the medical services really accessible to all people, must widen the limits of the scientific thought, must synthesize the experience and achievements of all schools and trends in medicine, and must optimize the relations between a doctor and a patient. It means a qualitatively different approach to the care for the health, an approach which respects and encompasses the common links between body, sensorium and environment.

The holistic approach gives its own answer to what such a change should and could be. It regards the health as a power of the harmony and the disease - as a disharmony, as a lack of harmony and peace.

From that point springs the basic target of the holistic medicine: to reveal and overcome the disharmony which had caused the pain or the disease. For the achievement of this aim without restrictions and prejudices, must be attracted all accessible diagnostic and therapeutic means, some of them orthodox, and others - not.

A guiding principle of the holistic approach in the treatment process is the minimum intervention with "appropriate medical preparations", in combination with the full set of sparing means (massages, exercises, diets, herbal treatment, and psychotherapy).

The patient is not only an object, but also an active and enjoying equal rights participant in his own treatment. While the treatment or the recuperation of the harmony is regarded as a direct result also of the change in the sensorium of the patient, of his inclinations, habits and beliefs. The overcoming of the psychological problems, the obliteration of the barriers of negative expectations, the positive thinking is conditions, leading to a change of the sensorium and recuperation of the harmonious order in the body and the soul of the patient.

As Marilynn Fergusson points out "the significance of the change in the sensorium in the process of treatment is probably the only and most important discovery of modern medical science". Not the simple physical change, but the state of the sensorium is the key to health.

The holistic approach regards man as an integral personality, and the body – as a dynamic system, a context, a field of energies in other fields and takes into consideration the cardinal role of his sensorium for his health.

We come to the place of the energy medicine. It accepts the holistic approach; it is guided by the principles of holistic medicine and is an inseparable part of it. In the core of the energy medicine lays the understanding for the human body as a sophisticated energy system, functioning in wider energy fields. The body is a process, a bio-electronic whirlpool. In it constantly flow, arrange, interact different energy streams. They change their force, direction, speed and are programmed to establish and maintain the divine harmony between body, mind and soul.

The energy is the basic constructive element of the body. It gives it life. It is also the drug in energy treatment. This drug applies to the physical energies and energy systems in the human organism. That is why the energy treatment is safe and natural. It harmonizes the vital systems by sparing them and safeguarding their integrity.

The energy treatment originates from the antiquity, but it is also one of the most precious possessions of contemporary medicine. It includes numerous healing practices from the past, as well as contemporary principles and techniques, leading to the establishment and maintenance of perfect health. It widens the capacities of modern medical diagnostics and therapy, but is not its alternative.

In what way shall establish, develop and enrich now and in future the energy treatment depends to a great extent also on the personal qualities, morality and the practical activity of the healers – bioenergy therapist.

I wouldn't like to depict here the ideal portrait of the bioenergy therapist. But he at all costs should be a personality with versatile culture, profound education and high moral virtues, be master of the secrets of his profession and should believe in the power of the human solidarity. By sending energy to his patients, he becomes part of their life, of their problems and joys. This high responsibility and mission he realizes with hard labor, knowledge, dedication and compassion.

I will take the liberty to make a comparison, notwithstanding, that every comparison simplifies in certain extent the problems. The role of the bioenergytherapeutist is very close to that of the stringer in music.

We could not admire the genial music of Bach, Beethoven, Mozart or the masterly performance of great musicians and orchestras, if in the music the man who tunes the musical instruments is missing. His name is not on the playbills of the concerts, he does not appear in front of the audience, but all, who write or play music know him, talk with respect about him and highly praise his work. He knows best every musical instrument, the materials from which it is made, the vibrations of the tons in it and with the almost invisible movements of his tools he is able to tune them in such a way, that they deeply move us with their miraculous sounds.

In order to be successful in his diagnostic and therapeutic practice, the bioenergytherapeutist must work along established and proven methods. These methods he should apply creatively by constantly improving his technique and widening his knowledge and skills, using the experience and achievements of other healers. He must be open for news in all fields of medicine, psychology, must work incessantly for his own improvement.

With his knowledge, experience and moral virtues he should induce trust in himself and in his practical activity, should predispose the patients to confide in him all their problems, doubts, to share with him the whole information which could be useful for the specification of the diagnosis and for the carrying out of the therapy.

The bioenergytherapeutist only wins, if he succeeds to inspire confidence in his patients, if he makes them active participants in the healing process. Quite often this is not easily achieved. In number of cases the patients have to change the stereotype of their life, their permanent habits and inclinations. If they are not convinced, that this is necessary, they will hardly want to do it. These changes are connected with the submission to a nutrition diet, with the introduction of new sleep and rest regimens, with giving up smoking and drinking and go further to a change in the thinking and the sensorium. Apart from this during the therapy the patient will have to do by himself exercises, cook himself herbal teas, organize his energy, and generally do a number of procedures by himself. And this is difficult, if he does not believe to a sufficient extent in his healer and in the methods for achieving of healing.

I depict separately the proposed diagnostic methods, along which I have been working already for twelve years, but considered together they form a diagnostic system. The use of each one of these methods separately or together with the others depends on the patient and his health problems. In some cases, in order to discover the disease it is enough to apply only one of the methods, while in others even after the complex use of all the methods still remain obscure problems. Then one should make additional tests and consult doctors.

The mastering and the application of the depicted methods give an opportunity to the healer to get sufficient information about the patient's condition. The energy signals, received through these methods allow revealing one complete instant picture of the psychic, mental and bodily condition of the patient.

With the help of techniques, which are an inseparable part of the methods, we can take out the pain from the body of the patient, organize his energy, and teach him to take care of himself, to build every day his energy field, to acquire self-confidence, security and power.

The diagnostics, performed according to the described mode, gives forthwith information to the doctor – specialist about the development of the disease in the very moment and directs him, if necessary, to specify the problem also with the means of the contemporary medical equipment.

It does not draw out the energy of the patient, on the contrary, it clears it from the vibrations of the diseases and frees it to flow and circulate in the body like pure healing and constructive energy.

In order to use the methods for diagnostics the bioenergytherapeutist should possess or acquire the following skills: to see the aura of the human body of living and non living matter; to be acquainted with the Chinese energy therapy. It is necessary for him to possess sufficient scientific knowledge in anatomy, physiology of man and his diseases.

The knowledge in radiesthesiology always could be used for measuring the intensity of the geomagnetic field and for determining of the appropriate place for work, sleep and rest.

It is of paramount importance to the healer to be healthy, to pay special care for this and to feel his organism in full harmony. The food, which he eats, should be ecologically clean. He should not drink alcohol, smoke tobacco and take other harmful for the organism ingredients and medicines.

Clad in garments of cotton fabrics in white and light colors the healer is feeling comfortably and is protected from the static electricity, which causes tension and disorganizes his biofield. The set hygienic rules are known: after every patient the hands should be washed with plenty of water. In the mornings and in the evenings the shower is obligatory.

The own system for relaxation, concentration and preparation for work of the bioenergytherapeutist makes him calm and confident in himself, ready to respond to any challenges of the patient such like lack of confidence, irritability, insolence, attempts to mislead him. He should not use the diagnostic methods, if he finds out even the slightest mistrust in the look or behavior of the patient. He should try to predispose him for this and to give him an opportunity to make his choice alone.

If the bioenergytherapeutist is a woman and she is pregnant, she should stop temporarily her work.

Conditions for implementation of the diagnostic procedure

Each of the energy diagnostic methods could be carried out in a medium size premise. The furnishing should be modest, without redundant furniture. A desk and a medical coach positioned in a neutral energy zone are enough. The walls by all means should be white. Most appropriate for them is the white latex dye, because it does not shine and absorbs the light.

The pictures and diagrams on the walls capture the look and can cause superposition of visual images, which may distract the bioenergytherapeutist from the objective and to mislead him.

There is no need of a TV set and computer in the working room. If you watch TV and work with a computer, you should wait for at least half an hour in order to start the diagnostic procedure.

The floor could be covered with marble or ceramic tiles in soft hues.

I recommend carrying out the diagnostics in day light. The luminescent light is harmful and causes deviations from the normal luminous image upon scanning.

The air in the room should be kept always fresh, with normal humidity and temperature.

It is of great importance for the results from the diagnostics not to change often the office and not to admit the presence of other persons, relatives of the patient or the healer.

The patient, according to his condition and to the methods for diagnostics which are applied is placed in upright, seated or lying position. His garments must be loose, not tight, not to press his body and must not be made of synthetic fabric. All decorations and objects made out of precious metals, which may cause deviations in the energy characteristic of man should be removed.

During the procedure the patient should be in the utmost possible psychic and physical stability, he should not talk, nor move his body. He should follow the instructions of the healer for building of his own energy protection.

He should not have drunk alcohol.

Description Of The Methodsfor Diagnostics

CHAPTER THREE

Bioenergy Diagnostics Of The Emotional Body

In the first years of my practice I carried out this diagnostic procedure by asking the patient to sit on a chair or to stand upright at a distance of two-three meters away from me. But this is not obligatory as the diagnostics could successfully be done even when the man being diagnosed is lying on a medical couch.

The following preparations must be made in order the emotional body to appear:

1) Teach the patient to do some very easy respiratory exercises, if up to this moment he has not been a subject to such diagnostics. Through his participation in the process he would detach his thought from the problems which engage him and would direct his attention to his breathing.

   If he had been practicing yoga, reyki, shiyazu, tay-chi etc. similar therapies and is acquainted with the energy centers of the human body I require from him also to fulfill even more sophisticated techniques...

2) Organize the energy of the patient. This technique I call also blowing of the channels. In fact the energy channels in the

human organism are concerned. From a distance of up to 30 centimeters I pass along the channels with my right hand.

3) After the described herein above had been performed by the patient and the operator, (let's call so my participation) I feel and see, that the field is spreading. After it grows up to 80 centimeters and I feel in my fingers that its density is sufficient I stand behind the head of my patient at a distance of half to one meter. My hands are upright and enfold the height of the created field. My eyes follow the contour of the physical body and after this concentrate in the tips of the toes. I close my eyes and I see the electric body of the person. I open my eyes and see him again the same. I transfer my look on the wall and there I can trace whether along the outer contour there are interruptions. The tracing could be made over the very body. I make the respective conclusions about the color of the electric counterpart and the density of its substance.

4) After I inhale air and close my eyes, the emotional body appears. Its colors are miraculously beautiful and hard to express in words. I recall one of my patients who eight years ago during the therapy were not answering my questions. In the end she told me that she had been seeing such beautiful colors that she kept silent from fear lest, if she answered my questions, this scene of Fairy could have disappeared. After the appearance of the emotional body it also has to be projected on the wall. Then at last my look returns above the body of the patient on the couch. My hands are constantly in front and palpate the created field.

The visual image of the lights of the electric counterpart and the emotional body, which I see around the patient and the vibrations that I feel in my palms and the fingers coincide. This gives me the opportunity to determine the size of the field, its vibrations and the character of the emitted waves.

From the luminous image of the emotional body and its size we can derive information about the emotional state of the patient and the organization of its energy. This characteristic constantly changes according to the change in the thoughts, feelings and reactions of the patient.

As the emotional body always appears with three consecutive light effects, this gives me the opportunity after respective analysis to add new details to the diagnostic characteristics.

Let's concentrate on the colors of the emotional body and the information which they give us.

Clear WHITE light springs from the people who are in mental, psychic and bodily balance. They have clearly defined aims and have succeeded to balance their "EGO" with the nature, their family and their work. Inherent to them are also such qualities like clarity of the thoughts and feelings and expressed striving for perfection. The appearance of a white light after the electric counterpart happens comparatively seldom. But if it appears, it is as if springing out of the body. Its waves are warm and elevate my hands; its vibrations are with high frequency I feel them like fine needles in my palms and fingers. In such a moment a man turns into a spring, into a generator of energy and could help other people and the nature with his thoughts and with propagation of this light. If he has been taught how to do it.

LIGHT PINK and LIGHT PURPLE light surrounds the people, who are in a state of harmony between thoughts and deeds, clear universal love, experience and spiritual elevation. It is inherent to conditions after meditation and relaxation. This color almost always appears after the therapy, which I apply: tenderness, beauty, love, warmth, relaxation and at the same time high self confidence. This is the light by which I can detect the people in love.

Its vibrations represent gentle fine pricks in the fingers and palms with medium frequency. The waves are limited to the length of the created color field and do not come out of its outlines while its size reaches up to 50 centimeters.

THE BLUE light from sky to deep blue is a proof of fidelity, of straight thoughts and selfless behavior, of great patience, self-respect, high labor efficiency. It is inherent to people, who have achieved harmony between their thoughts and deeds. Such people always seek the truth, believe in their capacities and are able to calm themselves and the others.

The vibrations of the blue light are soft, velvet, and the waves which they cause can reach up to two meters around the body.

THE GOLDEN YELLOW or GOLDEN light is a symbol of mind, knowledge and wisdom, clarity of the thoughts, compassion and readiness to help. It is joy from every venture, comprehension of the joy of life. I observe that almost all people, who meditate or pray, radiate with this light. During my participation in group meditations in India it was a real nirvana to me to observe the people in front of me to eradiate like lit candles. It fills even the emotional body of preachers, priests, rhetoricians. It is inherent also to vegetarians. Sometimes however it appears also from overloading of the thought (hyper energy).

This light is a sign of purification, no matter at what level it has been carried out - body, mental or spiritual.

Its vibrations are with mild frequency and the waves do not come out of the outlines of the created color field.

THE RED light is a symbol of life and love, of health, cheerfulness, vitality, joy. However, only in case it is within the outlines of the created field. If its vibrations are with very high frequency, they are transformed in exiting waves, which can fill the room. Such effect results in people, who are impatient, mistrustful, challenging, with expressed emotionality. Such people do not easily succumb to influence; they are selfish and are afraid of risky situations. They lead constant internal fight with the suggestions from outside and with their own controversial feelings. They often can cry and loose control over their emotions. The misbalance of their energy appears in the solar plexus. At bodily level they have problems with the digestive system. Gradually they acquire diseases, connected with the pancreas, gall bladder, colitis, low and high blood pressure. All people suffering from arthritis rheumatoides in acute form have red emotional body: they glow in red.

If the red light is dark and opaque at specific places, the patient should be advised to have blood tests, tests of the bone and the digestive systems.

A DARK RED (BORDEAUX) emotional body could appear if the man is a very active person. The appearance of this color means, that the thyroid gland works very actively and gives us a sign to examine the endocrine glands. The vibrations are with normal frequency and its field is about 20 centimeters.

A BROWN LIGHT in all hues appears around individuals with vague thoughts, who are in a state of some kind of difficulty, doubt, uncertainty and in most cases of inner rage. The damages at bodily level are in the vegetative nervous system, spondylosis of the spinal column, absolute misbalance of the energy, early stage of muscular dystrophy and some skin diseases.

A GREY ASHY light appears around people who have gone through stress, accidents, fear and different types of phobia, insufficient sleep, nightmares. Lately I see it quite often in children, who watch for a long time TV or play with a computer. It has very weak vibrations and a field of about 10 centimeters.

AN ORANGE light fills the emotional body of individuals with creative thoughts, ready for victory, for performance, of these, who are inclined for changes and have started in practice to accomplish them. Then the vibrations are concentrated and a field of about 80 centimeters is formed.

When a fruit reduction diet is initiated, after the fourth day the emotional body is filled with dense and very beautiful ORANGE light. This is a sign that a process of transformation of metabolism in the organism has started. Then its field is with dimensions of up to 20 centimeters.

If this light is slightly opaque and does not shine, does not glitter you must pay attention to the uro-genital system. Its vibrations are fine and with mild frequency. The field, which it forms, is up to 10 centimeters and this is a sign of the presence of fear, caused by shattered health.

THE PURPLE light. The dark violet or as they call it, the purple image, is seldom met. Individuals with such emotional body could be counted on the fingers of one hand. They are mentally and physically stable, self-confident just like this very light - symbol of power, tenacity and force. Actually the individuals who possess such strong emotional energy sometimes do not realize this. That is why they can injure themselves, their relatives and even the society.

It is very difficult to make an emotional characteristic of such people, one should be very cautious, because they are bearers of many controversial qualities. They can be with gentle character, love nature, and carry in themselves submission in front of its grandeur and the laws. But they could also fall in states of alienation from people, to become indifferent to the pains and sufferings of others, the asceticism to become their philosophy.

In persons with violet emotional body the diseases of the physical body are quite diverse. Some of them can be inherited while others to be characteristic for the respective geographic region and climate.

Its vibrations are intercepted as comparatively strong pricks but with normal frequency. The waves are even cool; the field keeps its constant dimensions.

THE AMETHYST light /purple to reddish-pink/ has very strong high frequency vibrations, its waves are warm and can enlarge the field of the emotional body up to 2 meters. This light is inherent to people, who have reached some stage of reconsidering things and growing wise, striving for simplicity and perfection.

THE GREEN light more seldom fills the emotional body although it is a symbol of growth, well being, and abundance. It accompanies persons, for whom the self control is natural fact. As soon as it appears I ask for the blood group and R?, because in most cases it is "0" with R? (-). Its vibrations are with medium frequency and its waves are slow and calm and do not come out of the field of the color image.

A BLACK light that has appeared after the electric counterpart is a sign of hatred, skepticism, evil. Sometimes it is a symptom that the patient is an object of a psychic attack.

It is also characteristic of people who are dealing with shady affairs, witchcraft, and charlatanry.

The light, bright colors which I see evoke in me the feeling of admiration of beauty. I call them divine, because they influence positively the people. The dark hues of the colors which have appeared have negative effect not only on the one who is their bearer, but also on the surrounding.

I have reached to the conclusion, that the individual color radiation of the emotional body of every man is different in different ages. It changes as a result of the groupings of the individual, the professional experience, and the changes in philosophy. In short, one can say this takes place along with the changes of personality.

CHAPTER FOUR

Bioenergy Diagnostics Of The Electric Counterpart

While the emotional body gives information first of all about the psychic state of the patient, with the help of the electric counterpart we can reveal the diseases of the physical body. With the help of this method, so to say, we study the history of the physical body, we discover its traumas suffered in the past and I receive knowledge about its present state.

Let us look at some of the luminous images of the electric counterpart which could be determined as typical for different diseases.

Diminished image of the electric counterpart

The head of the patient is outlined at the height of the solar plexus. The luminous image is intensely dove-grey. (Fig. 1)

For the first time I observed such an image of the electric counterpart in the Laboratory of Bioenergy in Plovdiv. It was of a man that had lost his child. He wanted to overcome this loss and was ready to do everything lest to take drugs. He had suffered from insomnia already for months on end.

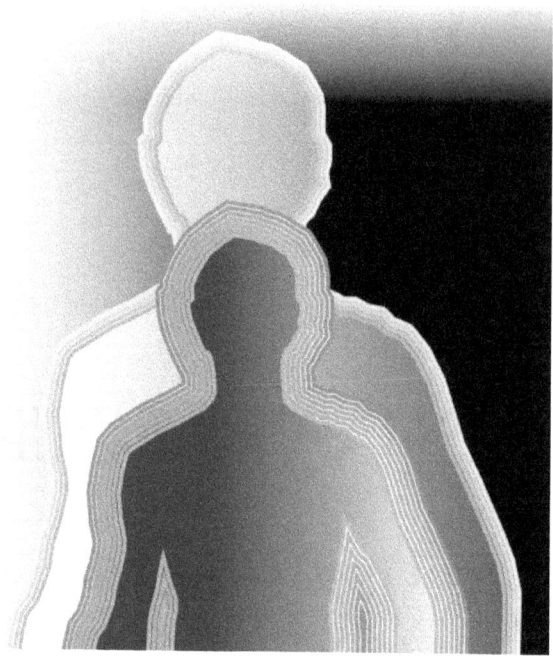

Fig. 1. The head of the patient is outlined at the height of the solar plexus. The luminous image is intensely dove-grey

I advised him to try to change some of his established everyday stereotypes. For example: changing his bed by moving for some time to sleep on a camp bed. Already on the next day he said that he had slept quietly the previous night. During the following therapeutic s?ances I taught him how to organize his energy, acquainted him with the net of Harman, taught him to discover the geo-pathogenic zones in his home and after three weeks his electric counterpart had absolutely normal image.

Such image of the electric counterpart has also the reticent, depressed people, who have been tormented, who are unsatisfied with their life. It appears also after stress from an accident, change of work, upon retiring and severance of the active sport and contesting activity or after loss of a next kin. I remember a woman from my native town, which after retiring kept waking up every morning at seven o'clock, stood in front of the window and watched how people went to work. She watched and cried.

The stress could diminish the dimensions of the electric counterpart for a long time and if measures for overcoming it are not undertaken diseases could appear mostly in the abdominal cavity, the bone and nervous systems.

Dislocation of the image of the electric counterpart behind and on the sides of the patient (Fig. 2)

The dislocation is most often from 20 to 50 centimeters to the left of the physical body. The colors are darker and denser in the region of the head. The light grey and dark pink to cherry color is dominating. Also bright effects like starlets sparkle in the region of the head.

Such an image appears in patients with neurological diseases of the type of hysteric neurosis, fright neurosis and depressive neurosis.

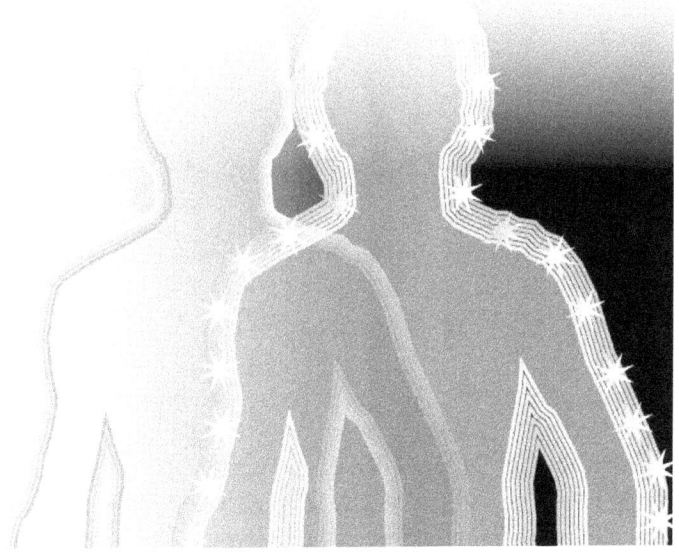

Fig. 2. Dislocation of the image of the electric counterpart behind and on the sides of the patient. The dislocation is most often from 20 to 50 centimeters.

An image of the electric counterpart, of which the head radiates with bright blue light (Fig. 3)

It could be compared to search-lights, crossing their lights in the frontal part of the head, above the fontanel. Such an image gives me information about previous insults.

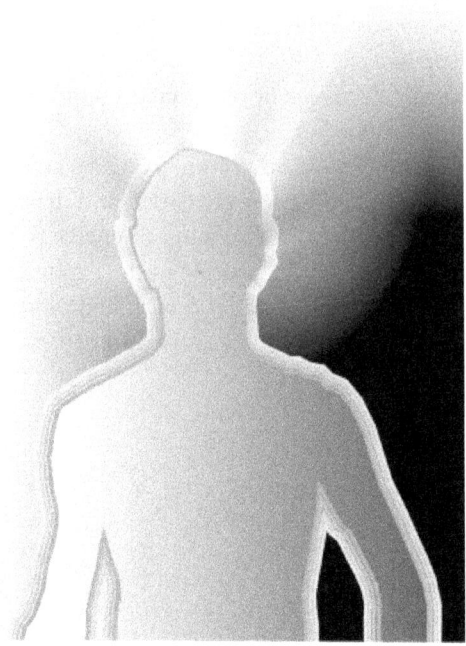

Fig. 3. Image of the electric counterpart, of which the head radiates with bright blue light

Image of the electric counterpart from bright-yellow to neon-green (Fig. 4)

It could sometimes appear even without an image around the head. It supplies information about radioactive radiation. I had seen such cases after the accident in Chernobyl. Five years after the tragedy I had the opportunity to observe children from the region of Chernobyl who had come to undergo treatment in Bulgaria. All of them had arthritis diseases and all had this image of the electric counterpart with the exception of the teacher, who accompanied them. When he understood from me, that he has not been irradiated he with happy cheer threw himself in the swimming pool with his clothes on and jumped like a child in the water.

In his book "Healing of the soul and the body" the Bulgarian herbalist Peter Dimkov has described a case of luminous image without head and explains that such a man can't live long.

I experienced the same case with a patient irradiated after the accident in Chernobyl whom I treated for a long time. Now he is feeling well but he had to change basically his way of life and nutrition. And the normal image of his electric counterpart around the head has recuperated already in the second month of treatment.

Image of the electric counterpart with changing density (Fig. 5)

The one half of the head is shining with bright blue light, while in the other part is formed a funnel with less expressed luminous and energy density. Such image gives us information about logo neurosis.

In the last years I traced many cases of acquired logo neurosis as a result of continuous watching of TV and playing computer electronic games.

Fig. 4. Image of the electric counterpart from bright-yellow to neon-green

Image of the electric counterpart in bright green neon light (Fig. 6)

The image is always with dimensions bigger than these of the patient. There is no graphite outline of the electric counterpart.

It submits information about serious disturbance of the immune system and AIDS.

I will not forget the first patient with such an image. It sprang in front of my eyes with some kind of unknown aggression. I recoiled and felt how the created field between me and the patient quickly increases and goes beyond the limits of the ordinary human biofield which is with dimensions of up to 180 centimeters.

Electric counterpart divided strictly into two parts along the length (Fig. 7)

Such an image appears after chemotherapy of breast cancer. I have observed several similar cases. Half of the image is black, the other half - yellow. In the process of treatment this picture is constantly changing and after curing appears the normal image of the electric counterpart.

Dark brown or black image of the electric counterpart with golden or orange frame (Fig. 8)

Such an image is a sure sign that the cancer disease has entered in its final stage. When a dark light is projected above certain organs this leads us to localization of the disease.

A dark light is observed above the lungs in cases of tuberculosis and above the liver in cases of cirrhosis.

During the treatment process I follow every day the changes of the aura and its bodies. This helps me to follow the oncoming changes and to prognosticate the course of the healing process.

Fig. 5. Image of the electric counterpart with changing density

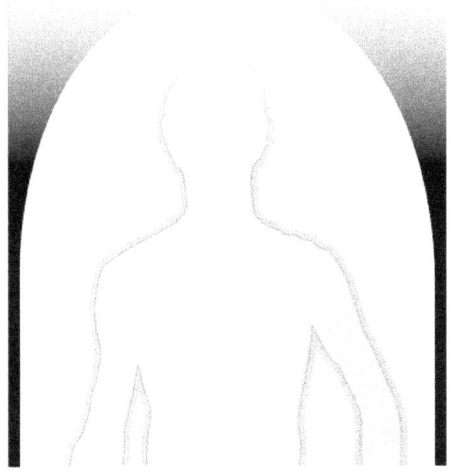

Fig. 6. Image of the electric counterpart in bright green neon light

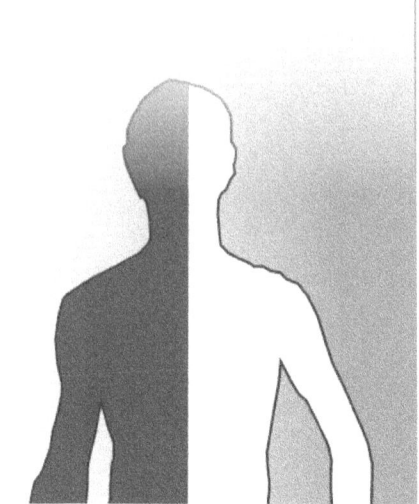

Fig. 7. Electric counterpart divided strictly into two parts along the length

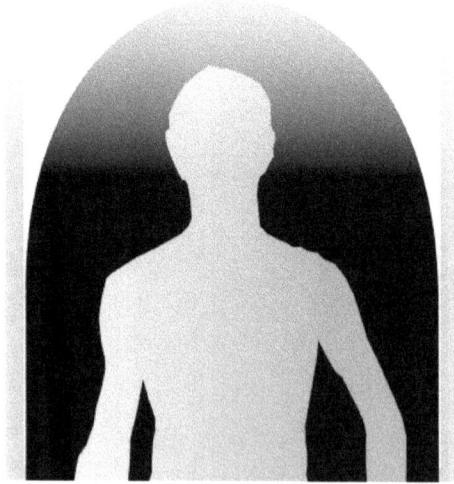

Fig. 8. Dark brown or black image of the electric counterpart with golden or orange frame

CHAPTER FIVE

Diagnostics Through Consecutive Remote Bioenergy Palpation

With this method I can penetrate in every cell, organ, and system of the human organism. I can draw information which after being processed and adjusted with the time of performance and the acquired experience, gives me the opportunity to make a characterization and to help for the discovery and treatment of many pathologic processes in the human organism.

For the implementation of this method are necessary first of all high extrasensory sensitivity, profound knowledge in anatomy and physiology of man and versatile information about the beginning, the causes and progress of the diseases in the human organism, as well as a number of other conditions. Very helpful is the knowledge of the methods of Chinese energy therapy and of the Eastern natural medicine.

The bigger the medical knowledge of the diagnostician is, the more the opportunities for effective implementation of the method grow stronger. Conscientiously I have challenged doctors - specialists to use the method to make a diagnosis of patients, suffering from diseases which are within their competence. After having mastered already the extrasensory sensitivity they manage quickly and precisely to find out with details the oncoming changes - through the registered vibrations in the tips of their fingers. A friend of mine —a gynecologist succeeded within seconds to determine the days of growth of the ovicell in the ovary of a patient already with the first try.

The implementation of the method is unthinkable without organizing and penetrating in the biofield of the patients.

THE ORGANISING of the biofield of the patient is a procedure, which increases its intensity. Technically I carry this out in the following manner:

The patient is lying on the medical couch.

I pass at a distance of 30-50 centimeters from his vertex and around it by using my both hands. My right hand moves from the fontanel aside and down on the left side of the patient and my left hand - respectively on his right side. In such a way I draw remote oval corresponding to the shape of the head and join my hands above the thyroid gland. I receive its vibrations and come out respectively from the outer side of the arms of the patient. In this way I distribute and organize the energy in this part of the body.

In the following position I distribute the energy from the solar plexus through the thymus gland and come out from the inner side of the arms of the patient. This passage normalizes and stabilizes not only the biofield, but also the blood pressure.

After this I make helical movements with both my hands around the head and join them above the thyroid gland around the shoulders above the heart, around the diaphragm, above the solar plexus, around and above the liver and the spleen and join them above the umbilicus.

From the umbilicus I pass to both sides of the pelvis and come out from the outer side of the legs, by using respectively both my hands.

The last passage is upwards: from the inner side of the legs to the coccyx (the tail bone). I continue to the umbilicus, then to the solar plexus. There I separate my hands and direct them to the acupuncture spot of the respective lung tips (apex pulmonis). I receive their vibrations and come out from the inner side of the hands of the patient – from both thumbs.

The described procedures clear the energy channels and increase the acceleration of the motive forces inside them. They improve the irrigation of the body by haematic and nervous way.

With the thus created clean and solid biofield, which enfolds the physical body of the patient and is with a range of about 50 centimeters, we can pass to the execution of the following procedure.

I perform the PENETRATION in the biofield of the patient through ORGANISING OF A REMOTE BIOENERGY CHANNEL. It is the main instrument, with which one works when using this method, the link between the energy systems of the diagnostician and the patient. I organize it with the efforts of my thought, by concentrating my attention on the object - the respective organ or vital system.

The remote bioenergy channel represents a highly organized bioenergy system, in which are included the signals sent by the operator to the patient and the answers received back. The diagnostician directs to the respective organ of the patient bioenergy impulses with specific aim: to obtain the necessary information in order to correct the already found out disharmony, to harmonize or stimulate its activity. According to the purpose the diagnostic channel has mainly information or mainly therapeutic function, or both functions simultaneously. As in the diagnostics the aim above all is to obtain sufficient information about the patient's condition I call the channel an information one.

How does it function?

I stand at starting position at a distance of up to 30 centimeters from the couch on which the patient is lying.

My left palm is inverted towards me in the field of my solar plexus.

I direct my right hand with the feeling that I am supplying energy that I am organizing a beam of light to the place, where my thought is concentrated until I am convinced that the remote bioenergy information channel has been already organized.

At the first moment I feel the interaction between the two biofields, mine and this of the patient and in the tips of my fingers is registered the pressure from their shrinking. A moment later I realize that the connection between the place of concentration and the tip of my fingers is accomplished. The formed information channel either raises or attracts my fingers in order to increase its intensity. My sensitivity increases as if an electronic device sets my position and determines the distance from which I can extract information. It constantly changes according to the present state of the organs and systems which I diagnose.

Upon penetration in concave organs for example in the esophagus, the stomach, the intestines, the organized bioenergy channel-beam fills them up by increasing its width and its volume. As a result of this the received information is sufficiently full and comprehensive.

How is actually realized the reception and the processing of the information received from the remote bioenergy channel-beam?

In the moment of organizing of the remote bioenergy channel in the tips of my fingers under the influence of the peripheral nervous system a sensation is felt for the increase of sensitivity, which gives me an information about touch and pressure, size and thermal radiation. With the penetration of the sent energy deeper and deeper in the observed organ the respective viscera-receptors are induced. The vibrations in the channel are increased or decreased and are dependent on the present state of the monitored organ or system. All changes registered in the tips of my fingers are received by the receptors and they, in their turn, re-transmit the information to the respective nerves. Reprocessed in the respective brain centre the received information gives me the opportunity to determine the state of the organ, its normal or pathologic condition.

The information bioenergy channel could be organized and directed to ourselves and could give us magnificent opportunity for auto diagnostics and auto therapy. It does not disintegrate upon passage through water. Diagnostics and therapy could be performed also to a man, who is in a tub filled with water.

Ivan Pavlov - the Russian physiologist for the first time noted that all nerve cells connected with the perception, transmission and analysis of the information for special type of irritants form an unified functional system called by him analyzer. Each analyzer has a peripheral section, made up of receptors. So for example the skin analyzer consists of receptors situated in the skin, from sensory nerves and a sensory zone in the pros encephalon.

In the human body every organ and system functions by pulsating, vibrating, swelling, shrinking, change its chemical composition. The application of this method gives more complete and more correct information about the physical condition of the patient, felt by ourselves, with our senses. Following each organ and each system at different time - in a normal state, in the periods of highest pressure or in a state of disease we acquire an experience with which we can prognosticate the development of the disease and take the respective measures for timely treatment. We can as well daily follow the course of treatment.

I shall try to treat the different types of sensitivity in a little bit more details.

Thermal sensitivity. The receptors for it are two types: for warm and for cold.

The receptors for warm react upon rise of the temperature in a range from 25 to 45 degrees Centigrade. They are situated unevenly in different parts of the body. Most of them are on the eyelids and the lips and in the tips of the fingers.

With them I feel the radiation from the surface of the body and in depth over inflammation processes, muscle spasms, at general rise the body temperature, above places with operative intervention, with high blood pressure, punctured wounds and others. Above the glands with internal secretion I feel thermal pulsations.

The receptors for cold are more numerous. They react with different frequency to impulses at temperature of the surrounding tissue in the range from 35 to 12 degrees Celsius.

Cold is the sensation, which I receive along the information channel in the tips of my fingers, in patients with low blood pressure, anemia, in cases of blocking of the energy from displacement and calcinations of discs of the spinal column, in cases of fractures of the bones, there where an organ has been extracted (spleen, kidney, gale bladder), above sectors of the liver, which do not function, above capsulated formations...

The information from the thermo receptors is conveyed along the ways of the somato-sensory analyzer through the nuclei of the thalamus to the posterior central plica of the cortex, the so called somato-sensory zone.

Pain sensitivity. The excitation of the receptors for pain sensitivity is conveyed through the spinal cord and the thalamus to the cortex of the brain (namely the second somato-sensory zone as well as in one zone from the frontal lobe). It switches in action also a number of reflex reactions.

Sometimes when I perform the method I feel strong dolorous prick in my fingers, which makes me even jump and withdraw my hand. My observations are that it appears when the patient feels pain at the very moment - with diseases of the type of neuritis, plexitis and most often at chondric cervical vertebrae. The osteochondrosis itself jams the energy in the central energy meridian. After passing along it with the organized remote bioenergy channel the energy is cleared out and its' natural flow is opened. In this case the operator feels a strong repelling effect.

In most cases my pain sensations coincide with those of the patient. But there are cases when I feel an old pain which the patient does not feel. It is like embers which could every moment burst into a fire and hurt. Most often this is felt in the region of the appendix, in the sex organs and the coccyx.

I feel the pain in the internal organs in cases of:

hypoxia (insufficient supply of the organ with oxygen);

distension of the organs, which have receptacle function: bladder, stomach, gall bladder;

strong and long contraction, spasm in the nonstriated muscle organs (stomach, intestines, bladder, blood vessels).

The inappropriate food may also cause pain.

In order to receive more detailed information from the internal organs I am helped by the viscera receptors. They are divided in several groups:

Mechanic receptors — situated in the walls of the blood vessels, of the organs of the digestive, respiratory and excretory system. The extension of the walls of these organs as a result of the change of the internal pressure in them leads to the excitation of these receptors. They are:

baro-ceptors or pressure ceptors for pressure;

stretch-ceptors, which determine the level of extension of the walls of the alveoli in the lungs, of the mucous membrane of the stomach and the bladder.

Through them I receive sensations with attracting and repelling effect.

The ATTRACTION is organized by motive forces inside the remote bioenergy channel which attract like a magnet the fingers of my hand to the respective place on the body and shorten the length of the channel. They are characteristic in cases of gastric pains and colic, hernia discalis, early phase of an organ atrophy, nephritic insufficiency, and cyst formations in the ovaries in women.

A REPELLING effect is obtained in cases when the patient has bronchitis and there is an extension of the walls of the alveoli. The repelling of my hand elongates the remote bioenergy channel which gives me the opportunity to extract the pain and reduce the pressure inside the organs. If the passage through the remote bioenergy channel continues via the anatomic duct of the system the pressure will be balanced and the patient shall feel improvement.

I feel a repelling effect also in the presence of a foreign body in the organism: concernments in the gall bladder and the kidneys, presence of metal prostheses, in hyper function of some organs, in fluid retention in the organism, in an intoxicated organism.

B) **Chemical ceptors** - they monitor the change of the concentration of a number of vitally important substances in the internal media.

I receive sensations in my fingers which tell me the chemical media of the substance. It is different for the different organs and for the respective diseases. It is expressed in cold, cool, warm and sticky moisture and dryness to a different extent.

I discern also the alkaline and acidic medium in the stomach of the patient, the specific acidity of alcohol and of some vegetable and animal food which had been taken by him.

For instance in patients suffering from diabetes that are undergoing insulin therapy I receive specific information in the tips of my fingers. It is dry but sticky and is emitted by the whole body and by all organs. In the ones suffering from rheumatism I have the sensation that from their whole body's cold moisture sticks to my hands.

OSCILLATIONS with different frequency fill the information channel according to the blood group of the patient, his present condition and the frequency of his brain waves. By the frequency of the brain oscillations I can direct my attention to the symptoms for foci of epilepsy and schizophrenia, for discovery of logo neurosis or localizing foreign bodies.

PRICKS. My fingers register light sporadic pricks which can pass in strong painful pricks.

In the events when during the passage above the energy channels of the organism an impenetrability of the energy is established I feel more regular pricks, accompanied with repelling. The recuperation of the energy flow is accompanied by a single strong dolorous prick.

I feel single pricks above varications, above chronic inflammations of the bronchi, infarctions, inflammation processes in the kidneys.

The long-lasting pricks are signal of exangia. They appear above the spleen, above the liver and sometimes - above the digestive system.

Sensation for the substance of the internal organs

This is a very specific sensation, which is acquired after specific practice. I receive information on the structure of the organ in a normal or pathologic condition. I feel in my hand humidity, glutinosity, dryness, density to a different extent, which helps me to assess the extent of the injury.

This phenomenon I call transformation of the substance. Upon diagnosing of strong inflammation and malignant processes like: tuberculosis, cancer of the lungs, cirrhoses of the liver, virus pneumonia, leukemia, diabetes it often happens that the operator extracts from the organism of the patient moisture and slime with different characteristics, depending on the progress of the disease. The sensation for glutinosity with different density and warmth is accompanied also by moisture on the tips of my fingers with different color – from grey to black and from light brown to dark. In such cases, in order to make precise diagnosis and to do a therapeutic procedure, I must often wash my hands with warm water and soap. This malignant energy could be extracted from the organism of the patient with a specific technique. But this is connected with great risks for the health of the diagnostician or the healer and he must possess a secure system for self-protection.

Upon coming into contact with a patient, who has a strong painful sensation, MOTIVE FORCES are organized in the remote bioenergy channel which leads my hand to the patient's organ and I start the diagnostics from there.

Thanks to the maximum concentration and mastering of the method the specialist could receive information on the human organism at a different level: cell, neuron, to penetrate in the sternum and to check the process of formation of the red blood cells etc.

Upon practicing of this method can be used also visual images - models of the respective system with the aim to concentrate the attention of the diagnostician and to achieve deeper penetration in it.

Sometimes in my sensorium are formed also visual images of situations in which the person had participated and so I can find out the causes of a stress or disease. For this purpose a very strong concentration is required as well as the presence of all conditions for implementation of the diagnostic process, mentioned at the beginning.

In the course of time when practicing the method I began to see the energy of the organized information channel-beam. This gives me additional information about the structure of its vibrations.

After I perform the diagnostics under this method I must organize the biofield of the patient.

CHAPTER SIX

Implementation Of The Method Per Systems

The remote bioenergy palpation of organs of the patient is carried out following certain sequence as I start always from the respiratory system. For me it is the key to the remaining systems of the human body.

Respiratory system

I concentrate myself on the hilus and organize two information channels with my two hands. My palms are turned inside one against the other at a distance of about one centimeter. In order to receive more precise information about the disposition and the tissue structure of the lungs, I pass several times up to the apexes of the lungs, after this downwards, following their lateral contour and again - through the hilus. (Fig. 9)

I palpate from a distance each separate bronchus and analyze the received information about the extent of its injury. (Fig. 10)

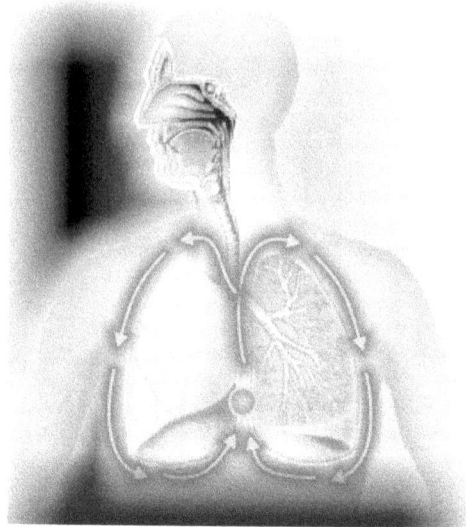

Fig. 9. I concentrate myself on the hilus and organize two information channels with my two hands. My palms are turned inside one against the other at a distance of about one centimeter

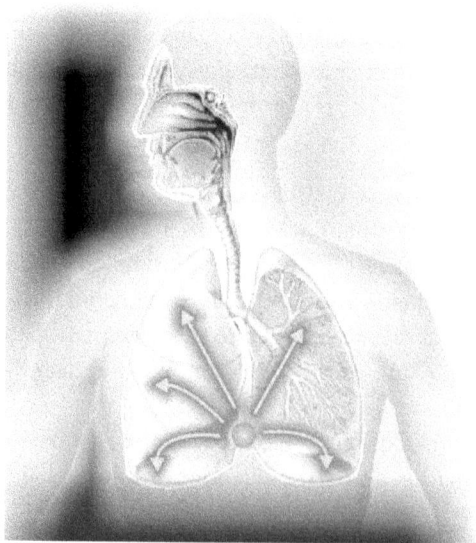

Fig. 10. I palpate from a distance each separate bronchus and analyze the received information about the extent of its injury

I concentrate myself again in starting position and at the moment of exhaling the air by the patient I penetrate mentally in the trachea. The information channel fills it and while passing upwards through it, until I come out from the nasal cavity, I register all changes, if any. (Fig. 11)

Upon repetition of these movements from 5 to 10 times appears as well the VISUAL COLOUR SCHEME of the disposition of the system, which I can project on the wall with my look.

The gathered information gives me an opportunity to determine:

the position of the lung: normal, slight descensus, peripheral consolidation, asymmetry of the lung apexes. For asymmetry of the lungs I judge from the asymmetry of the emitted signals by the apexes. According to the signal I can trace to what extent is the descensus, if any;

the thickness of the lungs, the breathing of the patient – thoracic or abdominal;

the tissue structure of the lungs and its grade of humidity;

eventual adhesions of the pleura and impaired innervations. In cases of adhesions of the pleura the information channel is interrupted in the moment, when my fingers pass over the very place. I feel coldness as well.

By the extent of the vibrations I determine whether the problem in the lungs is actual or from the past, whether it is a result of problems in the spinal column, osteochondrosis, and static deviations or is due to other reasons.

the reduction of the mobility of the diaphragm

sinusitis, polyps, problems with the upper respiratory tracts.

Also revealed are problems connected with the basic activity of the patient, the so called occupational injuries. In many cases for more precise assessment a long practicing of the method is required.

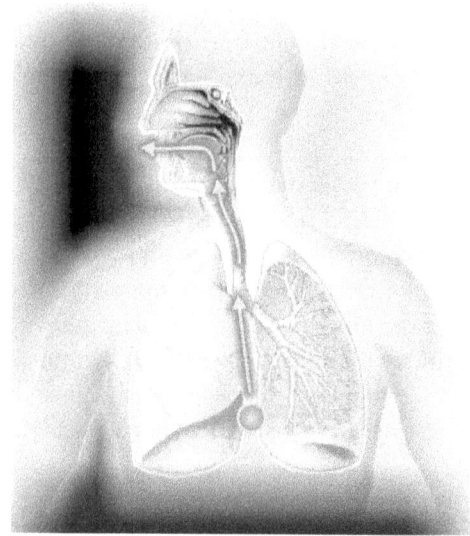

Fig. 11. I concentrate myself again in starting position and at the moment of exhaling the air by the patient I penetrate mentally in the trachea. The information channel fills it

By the vibration information received from the lungs I can differentiate the OCCUPATION OF THE PATIENT.

From teachers with practice longer than 5 years I get a sensation of light repulsion and coolness. The substance is drier than normal. Characteristic for them is the periodic breathing with increasing and decreasing frequency in depth. I have the sensation that their lung is veiled with thin septum, which hinders me to penetrate in depth.

The visual image has milk grey color.

Miners I detect by the sensation of stronger repelling effect. I have the feeling that the lung is narrowed and closed in a capsule with hard walls. I feel motive forces to the periphery of the lung. Characteristic in this case are the oscillations with slowed frequency and rare pricks, sometimes with pain sensations. I feel the breathing as a starting "sigh" with subsequent more shallow respiratory movements.

The visual image is dark to milk grey mist above the lungs.

Drivers, fitters and workers in petrol stations I tell apart by stabbing pain sensations with different intensity depending on the extent of injury.

The visual image is from dirty yellow to light brown.

Computer operators I identify by the high frequency oscillations in my fingers. At the same time I have the feeling of more energy with repelling effect.

The visual image is bright bluish grey light.

Workers in hot houses and in chemical laboratories cause in my hand sensations with attracting character with different frequency of the oscillations depending on the injuries.

The visual image is from intense yellow to dark brown.

Active smokers emit scratching vibrations on my fingers and heavy, sometimes sticky substance, by which could be distinguished also the smoking of marijuana and hashish, sniffing of coca and other narcotic substances.

The visual image is dark brown to black.

In a healthy man the vibrations are with normal frequency and I feel non intermittent energy signal.

The visual picture is beautiful bright green light.

Characteristics of the vibrations of lung diseases

In order to receive more complete and precise picture of the lungs status I penetrate through my imagination in every bronchus and feel the vibrations and its substance. I start first from the upper left bronchus and after that one by one I examine the lower left, upper right, middle and lower right bronchus.

Upon the diagnostics of each bronchus separately I receive information to what extent as a result of the disease have been affected the apexes and the posterior lobes of the upper lobe, as well as the uppermost segment of the lower lobe.

I move my hand very slowly; with the information signal I palpate every single millimeter of the bronchus. After this I can draw on a sheet of paper the calcinated place and determine its dimensions and form.

If the disease is currently in progress, upon diagnostics of the bronchi most often I receive a sensation for transformation of the substance.

Pulmonary tuberculosis (consumption)

In passed stage of the disease the vibrations represent strong stabbings in my fingers, which create visual picture of old, calcinated, localized foci, most often under the clavicle, in the form of a raceme.

The substance is dry. The extent of calcinations of the focus could be determined.

Luminous image - dark grey.

In the event the disease is running during the diagnostics, I immediately perceive the sensation for transformation of the substance at the place, where the process is localized. In some cases in the tips of my fingers I feel extracted moisture with different density (expectoration).

Luminous image - dark brown-reddish to black depending on the progress of the disease.

Cancer of the lungs

The vibrations in the tips of my fingers are high frequency and can determine the localization of the foci.

The sensations for the substance depend on the progress of the disease.

If the progress of the process is under control for the time being through medical treatment and chemotherapy I feel how cold grey moisture washes my fingers. In such cases the visual picture is with an electric counterpart with dark orange color, which comes out of its usual dimensions of 20 to 50 centimeters and has no graphite outline.

Should there be effusions in the lung during the treatment I feel the substance warm, aqueous and slimy. If the process is not curbed I feel thick, dark and sticky substance.

Visual picture: at the beginning – light brown, at the final stage- dark to black. The electric counterpart is without breasts and head.

Inflammation of the lungs

It is known that the PNEUMONIA occurs in inflammation of the lungs. In this disease the information bioenergy channel registers in the tips of my fingers warm high frequency pricks.

In case of retention of the expectoration I feel repelling signals and cool, thick and humid substance.

In the presence of festering foci in progress I feel again a repelling signal and extract in my fingers thick slippery substance but with higher temperature.

The visual picture above the impaired areas is dark grey thick mist which comes out of the outlines of the physical body, trespasses the graphite outline of the electric counterpart and in the peripheral regions is transformed in milk grey color.

The sensations for bronchopneumonia in children are other, different from these in the adults. The inflammation processes in the lungs of children could develop for several hours. This is because their lungs are small by volume and have weaker protective power than the lungs of the adults. The signals received in the tips of my fingers from the foci of the inflammation process are warm, with constantly changing frequency, accompanied by clear painful pricks in the acute stage of the disease and weaker in the process of healing.

I see the color of the healthy lung of a small child in pale green light. The actual color of the lung is pink.

Sensations for the structure of the lung:

It is aqueous, with variable temperature - from warm to cool and from normal aqueous to glutinous depending on the stage of the disease. In the early stages it is warmer, but when the disease recedes – it cools down.

At the places, where there is an inflammation of the lymphatic nodes I feel a repelling signal.

More frequent are the cases when at remote palpation of the bronchi cold moisture sticks to my hands.

The sensations depend also on the body temperature which in such cases is usually higher.

In this disease I always check the state of the kidneys and liver, I charge them with energy and I insufflate with the information channel their energy meridians.

Bronchial asthma

I discern allergic asthma which develops as a result of domestic and occupational environment of man - a result of toxic powders and harmful gasses, domestic dust, hair, climatic factors and others. My sensations are specific and they are most often expressed like transversal scratching as with a pin on the tips of my fingers.

In my practice I have had cases of asthma in different stages, with turgent lungs. Inside I feel heir substance as thick and stringy, with minimum humidity.

The visual picture is from dark grey to black. In such cases an orange emotional body appears, as a result of the experienced fear from suffocation.

After a choking fit the vibrations are like in a bee hive and my hands are attracted to the patient's body in the region of the lungs.

During passage at 20-30 centimeters above the lungs the dry substance of the mucous membrane is felt at places where it has atrophied.

Chronic bronchitis

This is one of the most frequent diseases with persistent cough. It occurs. Also after acute phase of an infection as persistent cough with expectoration, lasting at least three months in two consecutive years.

It is characterized with solid structure and well felt vibrations scratching my fingers.

The visual picture is from dark brown to black.

Digestive system

The cases with which I get in touch in my practice confirm every time the well known opinion about the effect of the feelings on man's health. I ask myself: What in our physical body reacts to every thought, to every feeling so compassionate, that gives us the opportunity to feel the thrills of the first love, of joy, of happiness, the pleasure from the delicious food? And in what way the negative emotions - anguish, sorrow; shrink our stomach in such a way, that we forget the feeling of hunger? The answer is – *the vagus nerve*. (Fig. 12)

It innervates almost all internal organs and they react in their own way to our emotions. If we study the vagus nerve in the anatomy textbooks we will become aware that we possess the most beautiful, most precious and perfect inner necklace. It responds to our gentle feelings with the most fine and high frequency vibrations, which can exist in nature. These vibrations give information about the hyper secretion of gastric juice and exuberant blood supply of the gastric wall.

And how the vagus reacts – the holy necklace when we feel unrequited love, when we are waiting for the next ordeal, when we are filled with sorrow, grief, impatience and anger? I feel how it shrinks into small balls and wriggles itself. The fine vibrations are transformed in thorns wishing to pierce my fingers and to revenge themselves by reducing the excretion of gastric juice, increasing the pulse rate, retaining food and waste matter in the body.

The digestive system is a peculiar mirror of our organism in which our health is reflected. It reacts to each mistake and ignorance which after this results in problems in the bone system, the heart, and the nerves. As soon as I palpate remotely the stomach of a man with my right hand I feel what feelings excite him, what food he eats, how much alcohol does he drink, how many pills and how many coffees he drinks daily. How does this happen? (Fig. 13)

I organize the remote bioenergy channel with starting point the buccal cavity. I penetrate inside along the tract of the digestive system through the pharynx, the esophagus, the stomach, the pylorus, the duodenum, the intestines, the colon and the rectum. I follow the gall and pancreatic juice by following their ducts to their portal in the duodenum.

Entering and penetrating inwards after the buccal cavity the information channel fills the esophagus. In the field created inside forces are organized which have spiral characteristic and thus I can palpate remotely every particle of the esophagus. For me the structure of the muscle fibers which build the very wall of the esophagus account for the nascence of these forces. They are spirally knotted as well as the remaining parts of the digestive tract.

The extracted information about the mucous membrane of the esophagus in healthy men convinces me that it has less humidity than the stomach while the warming in my fingers gives me information about its exuberant blood supply.

If you want to feel the vibrations of your own esophagus, you can take in turn too hot food and ice-cream. After this organize the information channel with your right hand towards you at a distance of up to 20 centimeters. The vibrations reproduced in the tips of your fingers will be unpleasant. They will be submitted to medium frequency dolorous needle pricks.

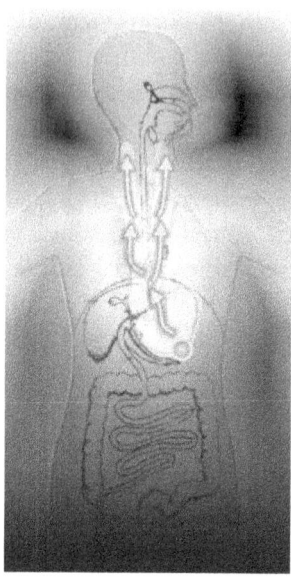

Fig. 12. What in our physical body reacts to every thought, to every feeling so compassionate, that gives us the opportunity to feel the thrills of the first love, of joy, of happiness, the pleasure from the delicious food?

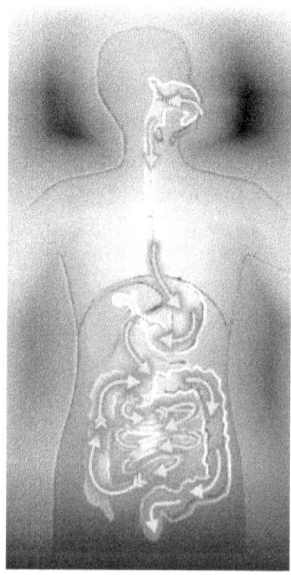

Fig. 13. I organize the remote bioenergy channel with starting point the buccal cavity.

I penetrate inside along the tract of the digestive system through the pharynx, the esophagus, the stomach, the pylorus, the duodenum, the intestines, the colon and the rectum

In the case of cirrhoses of the liver I receive from the walls of the mucous membrane of the esophagus warm dolorous pricks from areas with size up to 1 centimeter. This demonstrates that there is congestion in the veins of the esophagus.

So in an empty and without special naevi esophagus, the energy in the created information channel along the way downward to the stomach is stopped by the helical muscle. It opens only when the chewed food comes near to the cardia.

I pass further only in the case that some vibrations which have occurred do not stop my attention. I penetrate in the stomach. My hand curves to the left and makes several circular movements clockwise and extracts the tension in the information channel itself. Up to here already I have received information about the size of the stomach, its position and tissue structure, its state in the moment of examination.

I start careful palpation of the mucous membrane of the stomach until I get a sensation for its form and rugosity. The gathered signals give me the opportunity to determine the acidity or alkalinity of the gastric juice which for me is an important moment for the making of the nutrition program of the patient.

Probably here the question will come: Why the alkaline or acidic structure of the stomach juice? Is it not a temporary state depending on the taken food? Yes, food has its influence. But there is also heredity which determines mainly the more alkaline or more acidic reaction of the gastric juice in every specific person. I have made tests of families and always have found very similar picture of all vibrations and sensations inherited genetically from the respective parent.

The alkalinity or acidity is conveyed along the whole gastro-intestinal tract. The problems which are going to affect this system in the course of time depend on them. If acidity prevails this is an evidence of anxiety, which disturbs the normal vibrations of the vagus and the esophagus starts to burn and burn. I feel its mucous membrane dry and thirsty for moisture. In the created field high frequency dry vibrations scratch and penetrate my fingers. The information channel attracts my hand to the vagus which also needs help and as if ringing with high frequency rotations, calling for attention. It is in need of balancing the vibrations. Then I near my right hand above the place which has attracted me and I start to charge it with energy several minutes. I feel the nerve rugose and moniliform like green peas.

After the intervention of my right hand and the energy it had transferred to the vagus the vibrations of the nerve are normalized and I can continue my diagnostics deeper inside. Thus in most cases this diagnostic method is also like therapeutic method for balancing of the energy in the system and the blood pressure of the patient.

The increased acidity in the stomach is one of the causes for gastritis and ulcers. I can determine the stage of acidity and the size of wounds with the help of the information channel. It fills with high frequency specific and distinct vibrations and fixes precisely their size and position.

I feel the alkaline characteristics of the gastric juice in my fingers at remote penetration in the stomach. The picture on my way further is clear: more depositions on the walls of the intestines, less gastric juice, slime - reduced to minimum, information signal cool and slightly humid at places loosing its vibrations.

At normal characteristics of the gastric juice the information channel is filled with uninterrupted energy signal with vibrations with equal medium frequency. In such cases the passage along the system is achieved without special efforts and concentration.

In my practice I have seldom met people with mixed chemical characteristics. For example: from the stomach - vibrations for acidic substance, and from the intestines - for alkaline. This is one quite specific disease which causes on the one side of the digestive tract ulcers (I have in mind the stomach and the duodenum), and on other – constipation (obstipation) and difficult passage through the colon.

Another thing which should be obligatorily done is to receive information about the elasticity of the stomach muscles and their position, shape and size. I achieve this by palpating the stomach also on the outer side with the information channel. If I am content with this information I pass from the stomach to the pylorus and I receive sensation for its passage and elasticity. If with the information channel I cannot pass through the pylorus I perform several multi- circular movements with which I increase its energy and try again to feel the lumen and its vibrations. This means that the pylorus is constricted. It causes repelling hesitating vibrations with greater force. In such cases I have the feeling that I have placed my fingers on the apparatus for Kirliyans' photography. Always at greater concentration the energy in the information channel is increased and we enter the pylorus.

Passing through the pylorus I penetrate in the area with the form of a tuber called bulbus. I check carefully the walls of the bulbus and by the vibrations I detect ulcers, passed and active, if any. The healed wounds cause the sensation for attraction of my fingers and have low frequency vibrations. The active wounds repel my fingers and have higher frequency of the vibrations.

I continue downwards and with helical movements I palpate the walls of the duodenum. The sensations from it are with smaller mobility and stability at their location. The characteristics of the substance inside it and the vibrations, received from its walls, give me actual information about the status of the gall bladder and the pancreas and their activity. If I find it necessary I check them immediately by coming out firs along the duct of the system to the rectum. I remotely palpate the gall bladder, penetrate in it and pass in the gall duct and enter in the duodenum. This gives me information about the activity of the gall bladder. (Fig. 14)

There are cases in which I feel that for days through the gall duct has not passed any substance. It has turned into a dried up stream which already has changed the mucous membrane and with its high frequency pricks suggests me this. With the information received so far I continue downward and to the left, the information channel leads me to the intestines. Through them my hand, following the energy of the channel, makes zigzag irregular movements which follow their length. I penetrate in the colon.

Upon the passage through the intestines I feel smoother and cool structure, although in them the actual digestion is performed and because of this I always expect a warmer radiation. Sometimes I feel vibrations from retained non-digested food. If one is a vegetarian the radiation is colder and with fine touches on my fingers. If the taken food is meat warmer pricks with medium frequency come to me. In the presence of bacteria the sensation is as if I have put my hand in a bee hive and the bees hum and bite me at the same time.

In the case of normal function of the intestines I receive information which could be compared to the perception for dipping my hands in fine silk velvet. I disregard for a moment my thought and leave myself to my sensuality. Then my fingers seal the sensation that here inside rhythmically sway thousands of small crystal penduli, that millions microscopic little pincers in certain interval tweak the fine walls and they pulsate together with the velvet celia, creating one highly organized, fine and at the same time expelling flow. This sensation I can compare to a rare phenomenon by the sea coast: the starting of the breeze... The information channel is as if filled with this so delicate wind and in each instant it showers me with more and diverse sensations for the thin and fine structure of the muscle tissue. I feel the passage of the peristaltic waves whose interval I can always determine. The organized bioenergy channel causes also noises which I percept and I always utter: "Your intestines sing. This means everything is OK." In this song for me is coded special information connected with the central nervous system.

The passage to the colon is not always quick and easy. The information channel is interrupted if it finds problems, if there is retained food and that is why I return many times in order to feel its lumen.

Entering the colon I receive information about its internal volume. If the colon is filled with gasses its volume increases and fields of force are created repelling my hand.

I detect also depositions on the walls of the colon. Frequently I feel solid depositions with dry substance and repelling high frequency vibrations which outline their size.

Another variety is the depositions with medium hardness. These are the exangia, causing scratching on my fingers of medium-frequency vibrations. If they are in a process of exacerbation the vibrations pass in warm medium-frequency pricks, which are result of the congestion of venous blood and of the bleeding inspissations formed in the mucous membrane of the colon. Very often it is how I feel the hemorrhoids.

If at these places there are also bacterial infections I feel high frequency irregular oscillations, which at a definite interval pierce with light pain my fingers. Sometimes they even repel my hand.

Very important for the diagnostics is the assessment for the nature of the substance of the slime in the colon.

I receive different information about the exuberance and lubricity of the slime.

If there is a malignant disease in the colon the substance of the slime is thick, glutinous and a lot more exuberant than the normal. In the final stages of the disease appears the sensation for pultaceous matter and the information channel loses its energy.

I pass remotely in the rectum. Here I feel stronger and nonstriated muscles, more slime. My attention is directed also to the region of the link with the colon where hemorrhoids and polyps could be found.

Sensations in some diseases of the digestive system

Ulcer of the stomach and the duodenum

With the help of the remote bioenergy channel I can determine precisely the juxtaposition of the ulcers, their size and depth. The vibrations which I feel in my fingers from them are pricks with medium frequency, dependent on the changes which have taken place in the mucous membrane of the stomach and the duodenum. If there is a hemorrhage the vibrations are with higher frequency and temperature. Also emerges the feeling of transformation of the substance - an extraction of glutinous and warm sweat on the tips of my fingers.

The visual picture is in dark green.

I have come across cases of perforated ulcer. Due to the repelling vibrations with painful sensation in such cases it is impossible to organize the information channel.

Gastritis. Inflammation of the mucous membrane of the stomach

I feel in my fingers specific scratches with strictly defined medium frequency, coming from the whole stomach and the duodenum. In many cases appears also a feeling of specific warmth. Most frequently this is in cases of thickening of the gastric wall and stronger blood supply. I differentiate the chronic gastritis which is met in individuals for whom the acidic gastric medium is characteristic. The injury of the medium results most frequently from abuse of alcohol and coffee.

In rare cases I feel also cooler signals speaking of reduced blood supply and thinning of the mucous membrane.

Appendicitis. Inflammation of the appendix

At one of the seminaries years ago Doctor Kasarov asked me what my sensation of appendicitis is. In my childhood I suffered from chronic appendicitis. At that time without pondering I answered him: "What specific pulsation I feel from the glands with internal secretion, such I have also when is above this place of the human body. If I feel that the pulsation has changed its frequencies and the vibrations become higher and stronger, I send immediately my patients to a surgeon."

From my practical activity I realized also the link of the appendix with the genitals and mostly with the ovaries in women. Its increase is immediately conveyed to the right ovary and the same receives complications - cysts, inflammations, occlusion of the oviduct. The pain is so strong, that you can't understand what hurts you more. The vibrations which flow along the remote channel depend on the condition: from medium frequency with attracting effect to high frequency with repelling effect.

Inflammation of the intestines

The inflammatory changes of the mucous membrane of the intestines are felt along the whole duct of the remote bioenergy channel. The vibrations are high frequency, repelling, with stabbing pulsations from the passage of food inside the intestines. By the sensations which I receive I can determine the cause for inflammation.

in cases of cold – highly attracting cold information channel is formed, which leads my hand in the region of umbilicus;

in cases of allergy - warmer but repelling signal;

in cases of nervous factors - cold signal, repelling my hand from the impaired innervations;

in cases of bacteria and viruses – high frequency vibrations, pricking my fingers;

in cases of parasites – high frequency scratching in the fingers with repelling force.

Upon occlusion, paralysis and volvulus the vibrations are repelling from the whole abdominal part. I feel it like impenetrable wall. I direct immediately the patients to hospital.

Crone's disease

I have had until now 3 cases. One of my patients was a well known neurosurgeon from a seaside town in Bulgaria. I made a diagnosis through his clothes. Then he showed me his medical expertise and told me: "I know that there is no way out of it already. I was told that when you put your hands on the abdomen the pains of patients are gone. Advise me what am I to do by myself because I can't stay here." We worked out a nutrition program and I taught him how to charge with energy his vagus and his intestines.

I guessed because this was the second case. The first was an acquaintance of mine from whom I learned many details about this disease. He was not afraid of such type of treatment. We worked long and he felt relief. Once he came before a fit in order for me to see and feel what happens inside the intestines. I had placed my right hand in the region above the umbilicus and I was charging with energy when during several consecutive spasms his intestines were trying to gather in a ball near the stomach. I felt them empty, shrunk, without slime. All of a sudden it turned out that my hand was as if propped against his spinal column and the intestines retreated upward to the stomach. Then I put my left hand on his forehead while the right one remained in its previous place. Gradually the intestines started to regain their position.

Placing my left hand on the forehead and the right on the umbilicus gives me an opportunity to determine to what extent the innervations have been impaired. All embarrassed people suffer from impairment of the innervations. They can't express spontaneously neither their joy nor their sufferings but subdue them in themselves and so they cause themselves diseases of the stomach, the intestines, and the gall bladder.

In individuals of this type in the region of the solar plexus is frequently formed one intumescence. The information channel registers coarse grained structure of the vagus with stabbing repelling pricks.

The cause for the disease in the case depicted above was a strong stress from a tragedy that had taken place in front of his eyes. Looking through the window of his flat his look by chance stopped on a neighbor who was repairing his car in his garage. All of a sudden a blast was heard, the man jumped out of the garage with his clothes on fire. Only a couple of days after the incident the first symptoms of discomfort began and that is how this medical epicrisis was reached. And that are how this medical epicrisis was reached.

Colitis. Changes in the colon

I have described the changes in the passage along the colon.

Liver and gall

Diagnostics of the liver

It is also called the biggest gland in the organism of man.

I always compare it to a laboratory which not only must clean all taken human "mistakes" /alcohol, coffee/ but also create intermediate products and synthesis, necessary for the normal functioning of the whole organism.

I always pass to the diagnostics of this organ with a lot of love not only because it is unique and absolutely indispensable, but because it never hides information

I organize in my mind the remote bioenergy channel in order to determine the size of the liver, to discern its two lobes, to fix their position. I percept the vibrations of its flat frontal part, I pass in the hind and I try to palpate the portal vein. I feel warm; medium frequency pricks and penetrates inside through the created information field along the furcating of the portal vein. My fingers have received the vibrations in the channel and, watched from aside they perform many fine vibrating movements, by moving slowly and palpating every millimeter. I detect one fine and warm prick from the centre of each of the numerous structural lobes. I feel them like microscopic flowers, which immediately give me a signal that they are alive, that they exist, that they work.

If upon the passage I feel a repelling signal and solid substance this is a result from the filling in with retiform tissue of these lobes which already do not function. I can precisely determine the place and size of degenerated tissues. Sometimes I come also across bigger parts with solid septum in the form of a capsule, which have repelling and cooler vibrations. In most cases of application of this method a tension appears in the information bioenergy channel. Then I have to extract the energy to me and as I say, to clean it. I do the cleaning through my palms by strongly pressing my fingers against them. A crunching sound is heard. After balancing the energy in the channel I can penetrate immediately in the same place where I have been, but already with clearer signal and stronger sensitivity.

From my observations until now I have found out that in spite of the amazing regenerative abilities of the liver traces of past hepatitis are always found. A fine energy barrier is created after this disease which in the beginning of the palpation tries to stop my energy and does not give me an opportunity to penetrate further. I diagnose from a greater distance - 50-60 centimeters. My fingers follow the horizontal transverse movements of the energy which fills the information channel.

During a disease the injured cells increase the high frequency vibrations of the energy in the bioenergy channel.

In cases of *cirrhoses* of the liver I receive immediately sensation for transformation of the substance, especially if the disease is in a late stage. It is as if on my fingers sticks a thick glutinous liquid. The sensation in the final stage is for pulpous structure.

In people who drink alcohol I feel medium frequency vibrations and humid substance with expressed acidity depending on the injuries of the liver.

As a rule I consult my diagnosis with a doctor.

The gall bladder is a small sack which performs one very important and provident role. In it is gathered a thick liquid from the canaliculi of the liver, which fuse gradually in one channel. The gall bladder is a store place of the juice. If today, for example, we take more fats, then this reserve juice will help us to endure nutrition loading. This rescues us from great increase of the cholesterol.

Through the bioenergy remote and information channel I try to palpate the gall bladder with circular movements clockwise. I receive information about its position and I try to penetrate in the gall bladder. (Fig. 14)

I diagnose the bladder very carefully until I receive full information about its volume, passage, stage of humidity and about the presence of grit and concernments.

I direct my thought to the outlet of the gall bladder and the forces in the information channel fill the gall duct. I receive full information from it to its inlet in the duodenum. Along the line drawn by the information channel in the outlet of the bladder, I perceive the sensation whether it has the diskynesia. The conclusions which I draw for the functioning of the liver and the gall bladder take into consideration also the hours of take of food.

Upon the implementation of this method taking into account my practical activity I can formulate the following cases:

Normal function of the gall bladder. The vibrations in the information channel are with medium frequency, the substance is humid, and on the way to the duodenum from the beginning to the end the same uninterrupted energy signal pulsates.

Visual picture - light yellow.

Normal function of the liver and gall ducts with depositions in the intestines and colic. The energy in the information channel is with uninterrupted signal. Its acceleration is slowed after the penetration in the duodenum. The medium frequency vibrations pass into weaker ones with repelling or attracting effect according to the injuries.

Visual picture - fiery red.

The cause in most cases which I have witnessed is the late take of food in the evening after 9 o'clock, when the functions of this energy channel slow down and the food must stay up to 5 o'clock in the morning on its way to the intestine. When we add also the stress situations, the continuous nutrition trespasses of the laws nature and the hyperphagia we can regard the colic as pre announced result.

Recurved gall. During the passage from the gall bladder to the gall duct the information bioenergy channel determines the exact juxtaposition of the gall bladder and its duct and the remaining organs. According to the received sensations I make conclusions for the stage of the impaired innervations and contraction activity.

With the help of the remote bioenergy channel I can move downward the gathered liquid from the gall bladder, if there is still any.

Sensations for the substance in the duct to the duodenum:

- dry gall bladder and dry duct - the vibrations are attracting, scratching oscillations and with frequency depending on the stage of dryness;
- medium humid - vibrations with normal frequency and humidity;
- high humidity and acidity - it is felt in certain current conditions, when the taken food contains more alcohol and cholesterol. There are complaints from calor, epigastric burning. The sensations in these cases are accompanied by high frequency energy vibrations with warm signal and acute pricks.

Stone in the gall bladder. In the presence of a concrement in the gall bladder I receive repelling type of sensations, which follow its size.

In cases of small stones I feel strong repelling pricks in my fingers. For confirmation of the diagnosis in the presence of concrements I receive also one more signal. I see and hear clicking of small sparklings which fills the information channel.

Sometimes an unexpected movement of the energy in the information channel is created which leads me to different places along the way of the energy meridian of the gall bladder. Most frequently: in the left shoulder, in the region of the spleen and in the left thigh. The observation of this duct at the moment of diagnostics is important for the establishment of the causes of the disease. It can also be a disease of the plexitis or ishiatic type.

The visual picture is grey, thick fog, with static electricity.

In cases of operatively removed gall bladder I feel coolness above the extracted organ and perceive the vibrations of the energy counterpart of the extracted gall. When the patients who had undergone operation of the gall bladder follow strictly the recommended dietary regime the gall duct emits energy signal for normal humidity, witnessing constant outflow of bile.

The vibrations are with attracting or repelling signals according to the present state of the functioning of the system.

In cases of problems with the gall bladder changes in the breathing of the patient are also observed.

Pancreas

The pancreas is the biggest gland of the endocrine system which produces digestive juice. The effect of this juice is so great that it could successfully replace the functions of the remaining intestinal secretion. It not only disintegrates starch, proteins but also takes part in digestion of fats.

It is best to palpate remotely when the patient has turned his back towards me. Through the information channel I penetrate under the thoracic vertebrae, concentrate myself and supply energy in order to receive information about the position of the head of the pancreas, enveloped by the curve of the duodenum. Very quickly after this the information channel follows to the left the dimension of the gland and reaches to its tail, which borders with the spleen.

When there is a pathogenic process the penetration goes more quickly as there have been formed attracting forces which decrease the distance of the length of the channel. In this case in order to receive fuller and more precise information it is necessary to feed and concentrate more energy so that to increase the motive forces in the information channel from the diagnostician to the organ. After I palpate remotely and determine the size and the position of the gland I direct the information channel again to the head of the pancreas. From here I penetrate in draining channel, where gathers the secretion from the numerous gland moles. (Fig. 14) I pass along its whole length and receive information about the substance of the pancreatic juice and determine its quantity. I coordinate my finds for the functioning of the gland to the hours of diagnostics. Because they depend on the fact in what phase of the activity of the gland they have been made.

With normal functions of the gland the sensations are for uninterrupted information channel with fine medium frequency vibrations. I feel the substance with density of diluted yoghurt.

In the presence of chronic pancreatitis are registered cooler repelling energy vibrations and scarce quantity of pancreatic secretion which is with aqueous substance.

In cases of filling of the information bioenergy channel with high frequency vibrations, accompanied by harsh pricks and hyper secretion with thick substance, medical help should be sought immediately.

If there is a problem in the passage from the draining channel to the cuspidal part of the pancreas, where mainly the insulin and glukagon are produced, the forces in the information channel have attracting effect. In my fingers I receive the sensation for dry, glutinous substance.

If the function is totally impaired the signal cools and stops.

I direct the information channel back to the head of the pancreas and I penetrate in the pancreatic duct which is infused in the duodenum. (Fig. 14) In its end it is common with the gall duct. This remote bioenergy passage stimulates the activity of the pancreas and in most cases causes infusion of the ready secretion. The received information depends on the time when I must do the diagnostics and gives me full picture of the activity of the pancreas as well as of the passage of the channel to the duodenum. From here I always come out as I pass through the whole digestive channel. This helps for the melioration of the passage of the whole digestive system, gives it additional energy. Visual picture - green, if in normal status. In acute status – scarlet red.

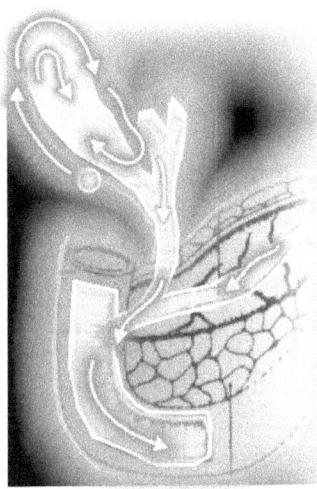

Fig. 14. After I palpate remotely and determine the size and the position of the gland I direct the information channel again to the head of the pancreas. From here I penetrate in draining channel, where gathers the secretion from the numerous gland moles. I pass along its whole length and receive information about the substance of the pancreatic juice and determine its quantity

Endocrine system

With each diagnostics of this system get I convinced that it possesses it own independent energy net. In this specific energy net every gland is a receiver and transmitter with specific frequency of the wave. Infortuitously the seven energy centers – the chakras in the eastern medicine coincide in their body situation with the endocrine glands.

The temperament and the emotional conditions of every single person influence and regulate all life processes in his organism.

The endocrine system - under the influence and in synchronism with the activity of the nervous system performs the main controlling functions in the human organism.

Part of the glands with internal secretion is under the influence of the hypothalamic-pituitary system in the brain. The governed by it glands are included in a common energy net. As the pituitary gland regulates the function of the thyroid gland, the suprarenal gland, cods and the ovaries it has energy way to every gland separately. The diagnostic bioenergy information channel gives us the opportunity to pass along these energy ways from the pituitary gland to every gland separately and to receive information for the passage of the system. It is passable when the information channel is filled with energy with strong uninterrupted vibrations, flooding in constant flow along the way from the pituitary gland to the ovary (cod).

The endocrine system reacts quickly to bioenergy signals. So some health problems could be neutralized with easiness.

In 1991 a girl to me came, complaining that after an operation of cysts of the ovaries she hasn't had periods for five months already. She was confused and discouraged. Then for the first time, as if someone was hinting me I passed bioenergetically along the endocrine system. I did this very quickly and accurately as if I had done it thousands of times. I was surprised by the force with which the formed information channel passed along the main energy way from the pituitary gland to the ovary. At about ten centimeters above the left ovary this flow was interrupted. But already with my first concentration and emitting of energy for recuperation of the information the link was established again. I felt a force pushing something that had stuck deep inside in the ovary. It warmed and started to frequent its vibrations. The same sensation received also the girl. After several days her menstrual cycle was recuperated.

I propose to you the opportunity to feel the strength of these links. They exist in the field of the electric counterpart. The whole endocrine system has its own strong vibration energy net. If you penetrate where it begins you could come out at the end with very useful information.

It is best for the patient to be seated on a high hard chair or to be standing but propped against a white wall.

I realize the diagnostic information bioenergy channel through my thought by concentrating the energy through the fingers of my right hand and directing it to the starting point in the head of the patient. I position my left hand slightly down, its middle finger pointing to the centre of the palm of the right hand. Thus in every moment the left hand can be involved in the diagnostics of the system. (Fig. 15)

I penetrate in the pituitary gland in my mind at full concentration from a distance of 50 to 100 centimeters from the patient. The penetration in every man is performed in a different way. In some people it is easier to pass through the fontanel downwards while in others I can affect it also from the energy centre of the very pituitary gland, the so called third eye.

Penetrating downwards to the vault of the hard palate I already start to receive also a color picture of the brain. The signals from the pituitary gland come first. I direct the information channel by turning it along a small circle clockwise and I feel the vibrations and the harder substance of the bones by which the pituitary gland is surrounded. The so called in medicine "sella turcica". A strong prick in my middle finger which passes in pulsations like these of a small heart gives me a sign that this is the pituitary gland.

I close my eyes, in order to see the bright pulsating light which I can fix already with open eyes on the white wall behind the patient. Here it shines and together with its pulsations, appears and contour of the brain. It shines with white light one octave lower than this of the pituitary gland. In the end of every pulsation this light attains light orange, sparkling hue. Because during this scanning I get tired very quickly I can not always specify with precision the next lights to appear after this. Sometimes there is an outline of light grey and another time – of golden yellow light.

In the luminous image I receive information for two lobes.

In its upper part I see a bright white light with many light bluish hues. It pulsates as a centre from which are dispersed fine wavelets with short luminous structure. They reach to the fixed image of the light from the brain about which I wrote herein above.

The lower part of the visual picture of the pituitary gland has a clear violet-pink light. I have fixed also cases when this place darkens to Bordeaux but the light is again clear.

86   Bio-Energy Diagnostics

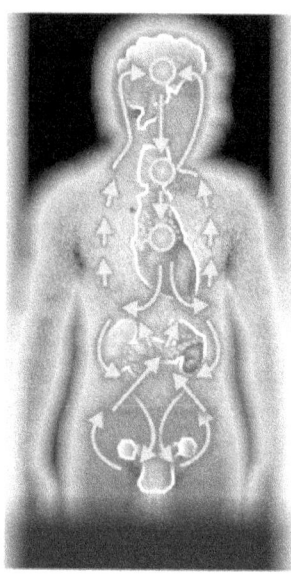

Fig. 15. I realize the diagnostic information bioenergy channel through my thought by concentrating the energy through the fingers of my right hand and directing it to the starting point in the head of the patient. I position my left hand slightly down, its middle finger pointing to the centre of the palm of the right hand

If there are problems with the pituitary gland the beautiful colors, the bright pulsating luminous globule and high frequency fine oscillations are replaced by a misty-grey outline with weak pulsation and vibrations. I had the opportunity to follow in the course of a month the emissions of the pituitary gland of a boy with proven inoperable brain tumor. He was treated with chemotherapy which had impaired also the pituitary gland. Each time when I concentrated to see how the things are going I had the feeling that his pituitary gland is just a float light, which slowly dies out. And so it happened, very quickly it died away. The boy also died.

In increased function of the pituitary gland in children it changes its colors at the time of the pulsations and could light in bright pink to red. In such cases I always send my patients to the doctor.

Sometimes it happens that I notice the emissions of the pituitary gland without the exertion of all these efforts about which I wrote herein above. If there is a problem this light appears spontaneously.

From the *pituitary gland* through the buccal cavity the diagnostic information channel leads me directly to the thyroid gland. I palpate its two lobes situated on both sides of the trachea and enveloping the larynx; I receive a sensation for its size. Penetrating in the thyroid gland itself I feel its grain structure humid, warm and exuberantly supplied with blood.

At normal functions of the thyroid gland pulsating distinct vibrations with the size of a lentil seed as if stick along the information channel on my fingers. This information is specific and corresponds to the pulsations of parathyroid glands.

In order to observe the visual picture of the thyroid gland the one from the pituitary gland should be cleared. Otherwise the result will be superposition of the pictures. At the same time you should not stop the flow of energy in the already created information channel. That is why still in the passage to the thyroid gland; I screw up hard my eyelids several times and open them with the feeling that I am ejecting the sealed image to the white wall. So it gradually begins to fade away and to move slightly upwards from the height of my look, the way of the information channel downwards is outlined as if with a golden band and the color image of the thyroid gland appears. It is from light to dark green mingled with blue. Such green we call petrol.

In cases of **hyper function of the thyroid gland** I feel an enlargement which is proportional to the progress of the disease and high frequency pulsating vibrations in my fingers.

In cases like this and other diseases the information bioenergy channel leads my hand to the problem and outlines functionally the impaired parts.

Above the warm nodes I feel high frequency vibrations accompanied by dolorous pricks which repel my fingers.

Above the cold nodes the information channel attracts my hand fixes the node and outlines its shape and size.

I have had two cases with malignant growths on the thyroid gland in which the light was so intense blue, that the only word with which I can determine it is "horrible" blue. It appeared in front and above the thyroid gland, swelling with every pulsation and reached the dimensions of a great apple. Sparks flashed in the periphery of the created image. I had the feeling that in the injured place is hidden dangerous destructive force.

After the received information from *the thyroid gland* I direct the information bioenergy channel to the thymus gland. It is bilobial organ and is situated above and in front of the heart.

The thymus provides me with information about the patient's age and his problems in crucial stages of life. I do not know how this happens, but it has as if coded in itself psychic information about stress situations in childhood and teenage which have been overcome. The information channel registers the changes that had occurred by stopping, it shuts up like a sewing machine. In such cases, in order to go on, it is necessary to concentrate myself even more. I leave my hand in the position, where to flow the vibratory duct had stopped and I start with circular movements to provoke the thymus. I suck to myself the caused tension and grind it between my palms. So it is purified, it revives and starts to emit slow, low frequency waves. Then it is possible in front of me to spring out the visual image of some situation from the life of the patient, which had later become a cause for his diseases and sufferings.

I will point out as an example one case from the last months. The patient is named Maria and is 26 years of age, with a lot of pains in the spinal column. This had gone on for several years. She had undergone many tests, but nevertheless the cause for her disease was not discovered. During the diagnostics my hand stopped on the thymus gland and I felt, even saw, how Maria falls from a slide at the age of five. I asked her immediately about that and she remembered in details what had happened. In this way it became clear where from originate all pains of Maria. In cases like this the thymus gland preserves the dimensions which it had had at the time of the accident.

It is known that with regard to the dimensions of the body the thymus is biggest at childbirth, doubles its dimensions during the puberty and after this gradually degenerates, as its functional tissue (T - lymphocytes) is replaced by fatty tissue. In childhood the thymus controls the development of the lymphoid tissue and the cellular immunity. When I pass through it in children up to the age of ten, I succeed to understand when puberty will start, whether it will be an early or late one. When the thymus is observed in the different ages of the different patients an experience for reception of specific information connected with the causes for the diseases is acquired.

The visual picture up to here is a continuation of the golden thread of the information channel, bolded in the place of the thymus so, that a cross is formed. At the time of nascence of images and pictures from different situations about which we talked herein above the light turns amethyst.

I continue downwards along the way of the endocrine system.

The information channel is divided after the end of the sternum in the region of the solar plexus. In the right side of the patient it begins to follow my left hand. I pass from the outside over the upper part of the pancreas. My right hand stops for an instant on the top of its left cuspidal part. Here I feel several impulsive jerks in my fingers and the information channel leads me quickly to *the suprarenal gland*. My left hand follows the right rounded part of the pancreas and after the curve receives the same impulsive impacts which lead it in the respective right suprarenal gland.

Everything happens very quickly, almost instantaneously, because the vibrations in the channel are strong. These are the energy corridors of the endocrine system about which I told you at the beginning. They have no connection with functional specifications for the regulation of the endocrine system.

The visual picture which I project on the wall behind the patient is distinctly expressed. The channels along which my two hands start are shaped very clearly, so is the pancreas. All they are in golden yellow light, which pulsating increases and decreases its glitter.

In the next instant I receive the pulsations of the suprarenal glands in my fingers by concentrating on the shape, the cortex and its core.

Provided the suprarenal glands function normally I receive medium frequency vibrations which push me to the ovaries or cads, depending on the sex of the patient.

In cases of sensation for strong frequency vibrations with repelling or attracting force accompanied by warming or coolness I direct myself again to every suprarenal gland separately. I penetrate in the cortex and in its core. In most cases if there is a problem, from here swirls one new direction of the channel – to some edema on the body or to a strong pain at certain place in the bone system.

Having in mind that the cortex excretes three types of hormones I follow every millimeter of it, in order to be able to feel and assess the respective emissions.

In arthritic diseases the vibrations are with low frequency. This demonstrates that the function of hormones regulating the metabolism of salts and water has been impaired.

In cases of changes and hypo function of the cortex of the suprarenal glands I receive low frequency vibrations and an information signal attracting my hand. In such cases I send my patients for consultation with a specialist and for hormonal laboratory tests.

The core of the suprarenal glands secretes the hormones epinephrine and norepinephrine. Individuals who have more epinephrine are in a state of constant impatience. In their eyes very frequently tears could be seen. If there is no one to analyze this state of theirs and to help them their nervous system comes very quickly to exhaustion.

I feel the high concentration of epinephrine as a warm repelling energy wave which leads me to the region of the heart. In such cases I pass remotely along the systematic circulation of the blood and control the blood pressure with special bioenergy technique. The electric counterpart of such people lights with greenish mist and the emotional body is intense red.

The energy connection suprarenal gland - ovary or cad is perpendicular and if it is not blocked I feel it with uninterrupted balanced signal which leads me straight to the point. In the ovary of the women are formed forces which lead my hand to the inverse lactiferous gland. With several circular movements I palpate remotely the breast. Having analyzed the received information from the vibrations I continue upwards along the acupuncture spots of the lactiferous glands. From there I penetrate in the channel of the triple suppurator and I come out of it in the region of the brow. This at first glance complex step helps for the reduction and eradication of the pains in hemicephalalgia.

The information channel assumes acceleration and so starts the diagnostics of the breast. With circular helical movements from the base of the breast to the nipple the volume of the mamma is palpated remotely very carefully. The presence of localized nodes, gouts or hardenings is established by the vibrations.

The inflamed lymphatic nodes have cool radiation and block the passability of the lymphatic channel in the armpits.

The mastopathy has emissions with different characteristics depending on the duration of the disease. If it is old the emissions are cool and with low frequency vibrations. If it has occurred recently they are warm and with high frequency vibrations.

About the sensations from the ovaries and the cods you will read in the description of the diagnostics of the reproductive system.

The passage through the endocrine system if it is passable takes not more than two to three minutes.

In the end of the penetration I receive a visual picture of the links and channels, along which I have passed.

Excretory system

The patient is seated on a high chair or is standing.

The diagnostics is carried out with two hands. Respectively my right hand organizes remote bioenergy channel with the left kidney of the patient, while the left - with the right. The distance is about 50-60 centimeters. My hands guided by my thought organize the circulation of the energy in two information channels. The right one moves clockwise and the left one – vice versa. So gradually and simultaneously I palpate both kidneys from a distance and establish their position. (Fig. 16)

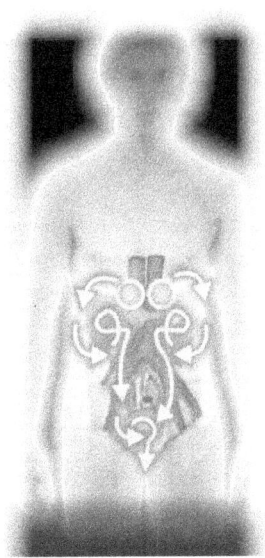

Fig. 16. My hands guided by my thought organize the circulation of the energy in two information channels. The right one moves clockwise and the left one – vice versa. So gradually and simultaneously I palpate both kidneys from a distance and establish their

position

The first sensation is for coolness, it comes from the fatty capsule, by which each kidney is enveloped. It turns out very frequently that the right kidney is slightly lower. This is connected with the position of the liver. Through the information channel I determine precisely the position of the kidneys and their size, after which I can draw them on a sheet of paper. From the dimensions and structure of each kidney I receive also the first information about its capacities to function normally.

The next, very important for the diagnostics moment, is the reception of information about the thickness of the cortex - I feel it rugose, smooth or cystic.

The rugous cortex emits medium frequency pricks, which register each microscopic fold.

In the cystic one the vibrations are with high frequency and I have the feeling that they withdraw me from the real cortex that is they create light repelling wave from the outline of the kidney. In this case the substance of the liquid in the cyst or cysts is also determined.

From the normal cortex I receive an uninterrupted information channel and the vibrations are with equal medium frequency of the wave.

I penetrate in the kidney through the hilus where the nephric artery enters in and the nephric vein comes out. I rotate the information channel from the upper part to the lower and I palpate the core in order to determine its substance.

I feel normal substance with medium frequency vibrations and normal thermal effect. In cases of substance with dry medium I feel multiple fine scratches on my fingers. This is a clear sign that the function of the kidney is reduced to minimum. At such internal substance the kidney itself has reduced dimensions from 3 to 5 centimeters. In such cases the sensation for reduced humidity of the substance is supplemented by sensation for its glutinosity.

In the basins of the kidney sometimes I detect also hard tough growths. I feel them like stranded small ships which create immediately a field with repelling vibrations. As if they don't let me touch them and have a closer look at them. My sensation for these thick structures is that they have much harder substance than the ones which are organized in the very kidney basin.

I concentrate my attention and through the information channel again I approach from the internal part of the hilus in order to penetrate in the kidney basin. I palpate its walls; I determine its size and substance.

In the presence of grits I feel very fine high frequency scratches in my fingers. I have the feeling that the grains of sand are between my fingers and I grind them with them. If small stones have been formed I feel from them repelling pricks, which as if glue the size of the stone itself. If the stones are bigger and harder I receive the already known cold repelling signal which outlines the dimensions.

I continue along the path of the system and I direct the information channel to the urethras which come out of the basin and enter in the bladder. I receive exact information on the trajectory of the urethras. It is important for the draining function of the kidneys and for its internal substance - dryness, humidity, and concrements. In all cases when I pass through the urethras and there are grits in them they are pushed downwards to the bladder by the energy in the information channel. The motive forces created in it clean the urethras from sand and small stones and the patient receives relief during the diagnostics.

I penetrate in the bladder. Here I join my two hands. The fingers of my left hand pass slightly downward and point the centre of my palm. In this position with the fingers of the right hand in front, I make diagnostics of the bladder through circular movements clockwise. I determine its size, fitness of its muscles, its position in the pelvis, the substance of its mucous membrane. If there are findings like stones their well known already vibrations stick on my fingers.

I come out through the draining duct of the patient and I continue with the two hands to move along the internal part of the legs to the final reflex spot on the nephric energy channel from the Chinese acupuncture system.

Sensations in the information channel in some nephric diseases

Inflammation of the renal basin (pyelitis)

Characteristic in this case are the medium frequency pricks in my fingers with attracting effect. Sensations for warmth and dryness are felt in the mucous membrane of the basin. The microbial presence is determined by the thermal emissions with specific frequency. I feel pricks with greater force from parts where the tissue has changed. If the disease is in advanced stage I feel also strong pricks as the pain could penetrate in my whole palm. In order to safeguard myself in such moments I stop and dip my hand in water. With the help of the vibrations is determined actually also the phase of the disease and the clinical picture of the patient. The substance also depends on the processes of retention of urine, high temperature and changes in the blood picture that has followed.

The visual picture which I get from the healthy kidney is in green, with the color of young grass and the volume of each kidney is outlined with light grey shining band.

From a pathologic kidney projects an opaque greenish-brown mist without precise outlines. The skin of the patient irradiates dark orange mist.

Nephrolythiasis

Concrements could be discovered both in the basin and in the core of the kidney. The created information channel sends me sensations of repelling type and such which distinguish the outlines of the concretion from the neighboring tissue. The harder the concretion the stronger and repelling vibrations it creates. By the force of the repelling reactions I judge whether the concretions are susceptible to breaking in a natural way. Such are for instance the phosphates and some mixed concrements.

Hydro nephrosis is found out when the renal concretion obstructs the basin. Then from the swollen as a balloon basin I feel warm repelling high frequency vibrations. My fingers are strongly moistened by the substance inside it. I direct the patient to a doctor specialist.

Contracted kidney - I feel the reduced size of the kidney and determine the stage of rigidity of the cortex through the vibrations — medium frequency soft pricks. Also – scanty glutinous substance which hints me that the blood pressure is not normal. I measure it and I establish renal hypertony.

In *the presence of glomerulo nephritis and renal insufficiency* I have the same sensations oscillating according to the extent of injury.

Cyst in the kidneys. The information channel in such cases outlines two images: the one follows the vibrations of the actual kidney without the tuber of the cyst, as if it passes below it. The other outlines the limits of the kidney and of the cyst. Medium frequency pricks which outline hollow spaces with aqueous substance - such is the characteristics of the second information signal.

In order to obtain more complete idea about the intercepted anomalies in the kidneys I check for swells on the limbs, pay attention to the eyes. I check remotely also the spleen. If the vibrations in it are high frequency I send the patient to be clinically tested.

Nephric diseases are various and diverse. Upon each noticed anomaly I send the patient to a doctor - urologist.

Genital system

Male genitals

In men I start the diagnostics from the kidneys. As it was described herein above I will continue from the moment when after the bladder, I penetrate in the prostate. (Fig. 17)

My passage through the prostate gland is realized through the central draining duct. The information channel fills it and registers the changes in the tissues; in its passage I receive also information about the substance of slime if any.

In normal function of the prostate gland the information remote channel has uninterrupted strong signal, with which I get out quickly of the body.

In acute inflammation of the prostate gland I feel high frequency vibrations with dolorous warm pricks. Then the information channel with difficulty passes through the draining duct of the prostate. Sometimes I have the feeling that this duct has disappeared and does not exist. This tells me that the inflammation has been caused by infection. With more efforts I can help the patient by organizing stronger remote channel with which I clear the way of the infection and the accumulated tension I pull outside. The pains decrease and in my hand I feel a strongly pricking my palms substance. Frequent are the cases when on the walls of the draining duct have stuck grit which scratch the tips of my fingers. Then the substance on the walls of the duct causes sensation for dryness and calor.

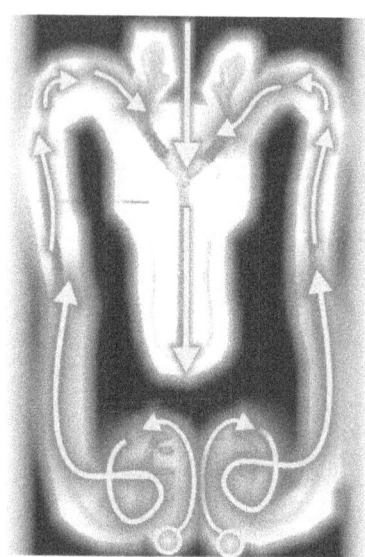

Fig. 17. In men I start the diagnostics from the kidneys. As it was described herein above I will continue from the moment when after the bladder, I penetrate in the prostate. My passage through the prostate gland is realized through the central draining duct.

Another way via which I pass in order to check the activity of the male genitals is the duct for excretion of seminal fluid through the prostate gland.

Remotely I penetrate in the cods, my left hand palpating accordingly the right cod and the right - the left one. Each one of my hands through the tips of my fingers creates its own information channel. My thought follows both channels simultaneously and registers through their vibrations the existing deviations from the normal status. With the tips my fingers I perform circular movements with small diameter in order to be able to more precisely and quickly determine the size and cover each millimeter of the very core of the testicles.

In cases of impairment of blood supply I feel warm low frequency, very weak pricks. In sections in seminal glands whereas the function is impaired weak, repelling vibrations touch me. I try to palpate the epididimus (appendix) which stands like a cap above the seminal glands there where the final maturation of the male gonocytes takes place. Then I can determine by the vibrations their quantity and quality. In a healthy man the vibrations are strong, medium frequency pulsations and shine pink.

I continue the movement by penetrating in the gonaducts. A strong energy information channel has been organized in them, which leads me to the gonocysts and the prostate gland. The information channel penetrates from both sides of the prostate inside it and is linked with the urethra. I continue along it and I come out while I pull out the created energetic tension in the channel. In such a way I clean it, I stimulate and revive the activity of the whole system. If there is a pain it is extracted by the motive forces inside the channel.

In the presence of *adenoma* of the prostate gland I receive strong repelling vibrations and in case there is a malignant growth – high frequency vibrations of attracting type which are with glutinous substance.

Female genitals

I palpate remotely the ovaries with both hands which create respective information channels. The movements are circular until the organized motive forces in the energy channel give me the outlines of the ovaries. I receive information about their size and position. I determine by their vibrations the changes in their cortex. (Fig. 18)

The cystic ovaries form high frequency vibrations in my fingers. So I receive the sensation for enlargement of the ovary. I receive also information about the substance, which is aqueous and creates tension in the organized channel. In such a case I must pull out the tension. If the substance is thicker, warm and glutinous, I direct the patient to a gynecologist because most probably these are "chocolate cysts", which I can't cure.

I penetrate in the core of the ovaries and determine the stage of progress of the ovicell. I can foresee the onset of the menstruation with precision of up to one day or two.

I penetrate in the gonaducts. The information channels outline strictly my way. I receive sensations for their permeability, for former or current diseases. I penetrate in the uterus. The fingers of my left hand move slightly downwards and point the middle of my right palm. I continue with circular movements to palpate the walls of the uterus with my right hand. The information channel is strong. This suggests also increase of the distance between my hand and the patient. I summarize the received information.

In *normal* state of the system a strong information channel with uninterrupted signal comes out with which I go out quickly through the neck of the uterus and vagina. The sensation for the substance is with normal humidity.

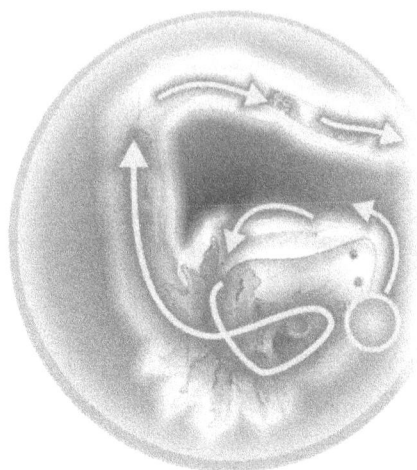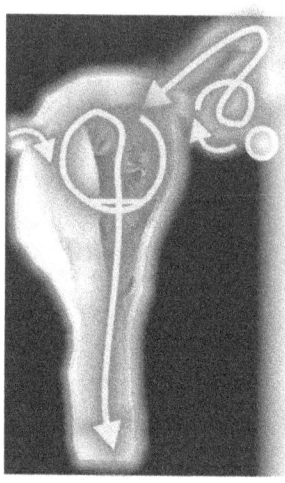

Fig. 18. I palpate remotely the ovaries with both hands which create respective information channels. The movements are circular until the organized motive forces in the energy channel give me the outlines of the ovaries

After abortion I receive information for dry mucous membrane of the uterus. It comes with medium frequency scratches in the fingers. I register the same vibrations also in the presence of wounds on the neck of the uterus as the information channel outlines precisely its dimensions.

In cases of other growths in the uterus of the myoma type I receive cool repelling vibrations of medium type, which outline the size and the form of the growth.

Sensations from attracting type, accompanied by high frequency pricks in my fingers are an indication of malignant growths.

Following the received information from the uterus I continue remotely to palpate the neck of the uterus and so I penetrate in the vagina. If the uterus is inverted or has distortion of the neck and the vagina to the left or to the right, I register it immediately. I determine also the substance inside the vagina. It could be slippery, aqueous or glutinous. As soon as I register the presence of glutinous substance I send my patients to gynecologist for additional tests.

The passage through the male and female reproductive system is done for two minutes. After this my visual analyzer has already fixed the system and looking to the white wall behind or on both sides of the patient I project a luminous scheme of the organs. This gives me the opportunity more carefully to analyze all details which are of interest to me.

For example a highly contrast picture with prevailing red light from the genitals appears when men and women abstain from sex in their active age.

Dark violet light emits the uterus due to some myoma growths. Then the contours of the energy counterpart of myoma growths are outlined with light grey light.

Here I want to summon your attention to a disease in the area of the ovaries caused by chronic appendicitis. From it suffer many young girls and women every month and this is determined as a plain ADNEXITIS. It is acquired as a result of an inosculation of the right ovary with the appendix. The pains during menstruation are very strong and acute. The diseased place emits warm vibration waves with attracting effect. This disease is liable to bioenergy treatment through which are ablated the inosculated organs and the normal function of the ovary is recovered.

With this method could be discovered in time ectopic gestation. A signal for this is the interruption of the information channel inside the gonaducts by a strong, warm and high frequency pulsating dotted substance.

Another opportunity of the method is that I can detect the fertilized gonacell still in the first week. I feel strong single prick which repels my hand. It is very different from other pathogenic vibrations, because it retains single and very fine high frequency. After the sensation of the characteristic single pricking appears also one microscopic white light at a distance of about ten centimeters from the uterus. This is the electric counterpart of the embryo.

The application of this method possesses not only diagnostic but also therapeutic effect. The duct of the system is checked, unblocked and cleaned with it. As the procedure lasts only several minutes, people think that miracles happen. But there are no miracles; everything is a result of precise dosage and direction of rational energy.

A young patient came to me with complaints that more than a year she had no menstruation. We made three procedures. I taught her how to organize her energy every day. After a week there was a result and six months since than everything with her periods is timely.

Mesaraic system and locomotorium

The state of the spinal column is of great importance for the normal activity of all vital systems. Even if only one of its inter cervical discs has a problem this affects unfavorably the whole locomotory activity of the human body, affects the self-confidence for soundness and force, changes the conduct of man.

The mesaraic system and the locomotorium is most significant prerequisite in order the body to be healthy and strong.

At the age of 30 I had the hernia discalis. The pains impaired my balance, made me feel unspeakable fear. It happened so that just then my elder daughter had to leave to study in another town. Up to that moment she had not separated from us and was worried, so was I. I mention this because I have read some healers say that the place of our wounds coincides with our psychological problems and the sacral area is the family. My sufferings were long-lasting; I could not do my habitual duties at home and in the office. But they helped me to become acquainted with my whole mesaraic system and locomotorium, its muscles and nerves, cartilage and tendons and to feel it like a motive force on which depend the activity of every organ and its respective system.

I start the diagnostics of this system always from the spinal column. I feel it like strongly vibrating antenna, which transmits its vibrations to every cell of the body and if the chain is broken start pains. Firs aches the place around the vertebra or the disc and after that sickens and the respective organ, which is innervated from there.

How to carry out remote bioenergy palpation of the spinal column

The patient could be standing or lying on the medical couch. At the beginning he is lying on his abdomen, and after that for testing of the sternum on his back.

I organize the remote bioenergy channel with my right hand and with the thought, that I must palpate the bones, joints, vertebra and discs to feel their structure, smoothness, cleanness and permeability. The bone system has its own energy net, its passable trail. Its structure possesses specific vibrations which I feel like finer needles, vibrating above the medium energy frequency.

I pass from the coccyx to the top of the head (Fig. 19). The energy in the information channel is organized in constant vibrations above medium frequency in the cases, where there is no problem. If the spinal column is crooked sideways - scoliosis, to the front - lordosis and backwards - kyphosis, the energy in the information channel follows precisely the bends. The vibrations direct me also to the causes for the distortions.

I concentrate myself again in the coccyx with both my hands; I check it and I pass along the sacral bone and to the respective coxofemoral (hip) joints. Respectively the motive forces inside in information channel also divide in two. (Fig. 19)

From the sacrum I receive medium frequency fine pricks. After passing above the joints I make several circular movements with which I clean the information channel and continue downwards to the thigh bone, the knee joint, the crural bone (tibia and fibula), the ankle and the bones of the footstep.

I analyze the received information.

I organize the remote bioenergy channel from the head (the fontanel) to the cervical vertebrae. I check them separately. After this in my mind I pull out the energy from the seventh cervical vertebra with my both hands to the shoulders and the shoulder joint. Above the shoulder blades I make several circular movements clockwise and pull back the information channel with the thought that I caress the humerus, the elbow joint, the spoke bone (radius) and elbow (ulnar) bone of the brachium, the wrist and I come out through the bones of the five fingers. (Fig. 19)

I diagnose **the sternum** by passing in front of the patient if he is standing. If he is on the medical couch he is lying on his back. With directed two hands I palpate the end of the sternum in the region of the solar plexus and I continue upward to the collar bone (clavicle). Here I divide my hands and pass over the ribs – respectively on the left and right side. The vibrations received from this bone are very fine and direct me to the substance inside the bone. And if I want, with additional concentration I can detect the regions of formation of the red blood cells. (Fig. 20)

The visual picture in a healthy man – the sternum radiates in light green and the ribs - in light yellow.

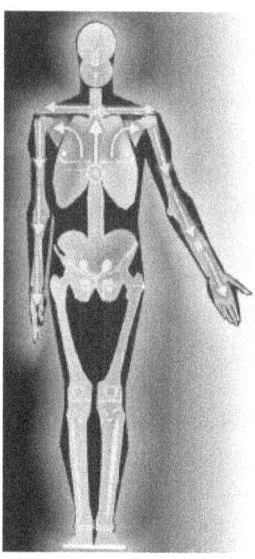

Fig. 19. I concentrate myself again in the coccyx with both my hands; I check it and I pass along the sacral bone and to the respective coxofemoral (hip) joints. Respectively the motive forces inside in information channel also divide in two

From the created vibration forces in the information channel its energy stops there, where there are problems: displaced disc, incremented calcium, distortions, arthrosis of the joints, etc. By this method of diagnostics I detect also the invisible with unaided eye distortions of the spinal column and impairments of its mesaraic and locomotorium function. I feel the substance inside the joints, also the interruptions along the neuro-reflex way.

In the case of normal osteal system I register constant flow of energy in the information channel with vibrations above medium frequency. The visual picture shines in light yellow already in the first passage above the spinal column. In the presence of deviations from the normal structure they shine with hues from lighter to darker bluish.

The visual picture correlates /coincides/ with the sensations in my fingers.

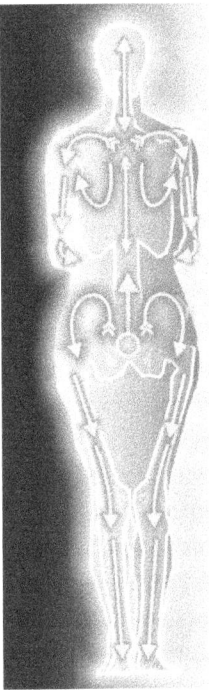

Fig. 20. With directed two hands I palpate the end of the sternum in the region of the solar plexus and I continue upward to the collar bone (clavicle). Here I divide my hands and pass over the ribs – respectively on the left and right side. The vibrations received from this bone are very fine and direct me to the substance inside the bone. And if I want, with additional concentration I can detect the regions of formation of the red blood cells

Sensations in some diseases of the mesaraic system and the locomotorium

Injury of intracervical disc

From my practice I found out that this disease occurs in every age. Most frequently modify and ossify the discs of the cervical joints between the second and the third and between the sixth and the seventh cervical vertebrae, the discs between 6 and 7 thoracic vertebrae in the lumbar segment and of L-4 and L-5 in the sacral segment L5 - S1. This causes interruption of the information channel above the impaired places. I can find out the number of the most severely damaged discs. If the received vibrations are low frequency needle scratches in the tips of my fingers, it is a question of depositions with high hardness.

The visual picture received from these parts is shining light-blue light. In the presence of growths with smoother and thinner structure I feel medium frequency vibrations and start their eradication through charging with energy. If the patient has dolorous sensations I receive them in my fingers like thermal pricks with different frequency.

The disease of Behterev causes repelling vibrations most frequently in the cervical and the lumbar segment of the spinal column.

The luminous picture is grey mist with numerous twinkling starlets which fill the luminous band, outlining the spinal column.

In the presence of *arthritis* of the coxofemoral joint I feel dry high frequency needle pricks, sometimes with pain accent.

The luminous picture is dark grey mist around the joints.

The hernia discalis causes warm high frequency vibrating sensation in my hands. The information channel and its motive forces lead my hand to the place of the injured nerve.

The luminous picture is light green dim light.

Hernia discalis – high frequency needle vibrations repel lightly my fingers and outline a diameter of about ten centimeters above the injured place. After slight movement of the patient I receive dolorous prick, coinciding with his pain.

The visual picture is like pink mist above the injured place.

Arthrosis of the joints

If the arthrosis in the moment of the diagnostics is not acute and dolorous I feel dry, gentle, high frequency pricks with cool repelling character, with which are fixed the precise places of the injuries. If the joint is inflamed and the patient feels pain I find the precise place, from where the pain comes out. I feel it like warm jet which lifts my hand. I pull it out with circular movements - to its reduction and stop. The pain itself increases the tension in the information channel and as if it flows out from it.

The visual picture is grey mist with electric charge in which twinkle millions of micro particles.

In the presence of *bone fractures* the information channel fills with cool, repelling energy precisely above the place of fracture. If the fracture is accompanied as is in most cases with haematic edema I concentrate myself in order to separate the two superposing pieces of information: the cold – from the bone and the warm – from the edema.

If there is a punctured rib I feel strong concentrated pricks, corresponding to the pains of the patient and motive forces which carry my hand along the whole length of the rib. In such cases I pull out the pain via the pathway of the motive energy force. I follow the electric counterpart if it is interrupted of the wounded place – the bone is broken.

In the presence of *tumors in the bones* the first sensation is for interruption of the information channel. It is accompanied with clicking which is similar to this of the demagnetization of the computer monitor in the moment of switching off. The electric counterpart is outlined with opaque grey-bluish mist, which covers specific space above the place of the tumor. The vibrations are high frequency; magnetism is felt in the tips of my fingers. There is no clear picture of the bone itself, everything is disintegrated in a common magnetic field which enlarges. One should immediately stop work, to clear the visual picture and to take measures for cleaning of the energy in working room.

Rheumatism. Rheumatoid arthritis. In the presence of these diseases I feel something so specific like vibrations, like energy, which is cannot be mistaken with anything else. Sometimes patients tell me: "I made tests, I got no rheumatism". But I see it before even I have passed along the skeleton through the method, which I related herein above. The information channel fills with uninterrupted medium frequency vibrations, which as if etch my fingers. They are covered with cold moisture. Immediately I receive also red visual picture of the bone along which I have just passed. These diseases have so strong, filling, enveloping and penetrating energy, as if the patients transfer constantly their pains to those around. I have the sensation, that solely with it is possible infection from a distance as a result of the influencing of the emitted energy of the sick. If an adult man, suffering from rheumatism holds in his hands a small baby, it will be restless, and probably it could cry and twists. This disease spreads more and more and encompasses great part of the young people. I discern its energy even when walking on the street and feel the vibrations of coming towards me diseased people but sometimes suspecting nothing persons. I am sure that in future I could tell something more about this disease.

Nervous system

The diagnostics of nervous system requires concentration. The diagnostic bioenergy channel I feel like a fine laser beam, which I can direct and with which I can penetrate in every cell.

The first procedure which I perform is the passage along the centrifugal ways from the tips of the five fingers of my extremities to motory centers in the cortex of the brain.

The patient is lying on his back. With my left hand I organize the energy remote energy channel above his head and fix the motory fields of the five fingers of the left hand of the patient. They are in the right part of the brain cortex. With my right hand I perceive separately the vibrations from the tips of the five fingers of his left hand and I pass upward through the truncus cerebri aside from the fore medium slot of the medulla oblongata, through which passes the pyramidal tract. At this place cross also the exodic pathways of the five fingers of the left hand to the right part of the motory cortex. The information channel fills with equal high frequency fine oscillations if the organism is healthy. So I pass also from the fingers of the right hand (and respectively both legs) having in mind that this nervous pathway is crossed. Which means that in order to receive information from the fingers of the right hand I have to fix the centers of the five fingers in the respective regions of the motory cortex of the brain? The difference could be felt if one passes along the same pathway of patients with juvenile cerebral paresis. In this disease the energy in the channel is interrupted but this procedure stimulates the nervous system of the patient and his movements become smoother. This also refers to the remaining diseases of the nervous system linked with the motory functions. If the diseases are subject to therapy: with hernia discalis, inflammation of the ishiatic nerve, nocturnal enuresis of children and other every day I check the recuperation of the pyramidal tract.

Diagnostics of the cerebellum

The technique of diagnostics of the cerebellum is similar to this for the diagnostics of the lungs. (Fig. 21)

The patient is turned with his back towards me. I concentrate my attention to the posterior cranial fossa and especially to the lower part of the so called "worm", where knit the two hemispheres of the cerebellum. As every hemisphere is linked with the trunk through three pairs of nerve bundles (cerebellar peduncles): superior (brachium conjunctivum), middle (brachium pontis) and inferior (restiform body) I program myself to palpate the two inferior peduncles. In the first moment both my hands are pointing this place but after they feel the organizing of the energy in the remote bioenergy channel they separate with circular movements respectively to the lower part of the two hemispheres. Here passes part of the pathway of the deep sensitivity and always comes out tension in the channel. I pull it lightly from both sides with my hands and rub it in my palms. I check in the same way the middle and upper peduncles. If everything is normal the information channel vibrates with medium frequency vibrations, which have specific frequency. As if the information channel is filled with static electricity.

From the same starting position I palpate remotely and from the outside the two hemispheres of the cerebellum and I penetrate along the trunk. This procedure again is in its way cleaning of the tension, but has also expressed curative effect in diseases of the vestibular apparatus, high blood pressure, problems with osteochondrosis of the cervical and other vertebrae of the spinal column and all diseases of the nervous system. In most cases the problems in the discs of the cervical vertebrae interrupt the energy information from the trunk and create short, transversal vibration lines. This is a sign for disturbances of the co-ordination of the movements and suggests additional clarification of the cause. In such cases the energy in the information channels takes accelerating passage to the injured organ or disc.

108   Bio-Energy Diagnostics

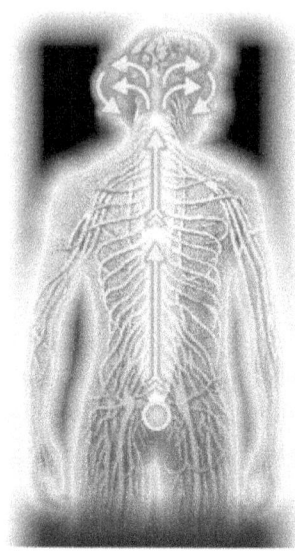

Fig. 21. The patient is turned with his back towards me. I concentrate my attention to the posterior cranial fossa and especially to the lower part of the so called "worm", where knit the two hemispheres of the cerebellum

With assistant professor Vera Tocheva in the Laboratory of Bioenergy in Plovdiv in 1990 and 1991 we observed that when I pass every day in the described way in the cerebellum of children, suffering from juvenile cerebral paresis the co-ordination of their movements is clearly approved. In the beginning of the procedure I feel pricks with low frequency vibrations, but not from all parts of the nerve bundles. After 15 days therapy the frequency of the vibrations emitted from the cerebellar peduncles is increased.

During the diagnostics of the cerebellum an oval light yellow picture of the whole head is obtained. The cerebellum shines with even brighter and shining yellow light. When I project the fixed image on the white wall I wait a little and then appear white pulsating specks, corresponding to the vibrations which my fingers have felt.

Remote bioenergy palpation of the vagus

I concentrate in the vegetative plexi of the abdomen, I pass upward along the pathway of the vagus to the esophagus, and I climb upward along the carotid artery and enter its cores, situated in the medulla oblongata. (Fig. 12 and 22) The cores always have pulsating effect on my fingers. So, during the sensation in the core of the high frequency vibrations with thermal effect, the inflammatory process could be prevented on time by directing the patient to the respective clinic and specialists. If from the core of the vagus are felt repelling vibrations one must check for eventual compressing effect of neoplasm on the core. A consultation with specialists is necessary also in cases when a pulsating prick of the core is not felt.

In cases of diseases of the organs, innervated by the vagus, structural changes of the nerve near them are registered with repelling energy, corresponding to the thickening of its fibers.

Very frequently in problems in the organs, which are innervated by the vagus (pharynx, larynx, lungs, heart, esophagus, stomach) I pass along the centrifugal pathways to the core of the vagus. A quick mitigation of their sufferings is achieved because every disease reflects also in the respective nerves. For the first time I adjusted myself to pass through the vagus to its cores in the medulla oblongata when I had to help a next kin and no medical help was available nearby. Before I made up my mind I used almost everything that I knew, but there was no result. He suffered from asthma...The positive result of the bioenergy passage along the vagus came so quickly that after this for days on end I reproduced my sensations, in order to convince myself, that I have done it. The dyspnoea stopped as a miracle and after this six months did not recur. Already eight years I use this technique. A mollification of the pain comes always; the breathing and the blood pressure are normalized.

By using this method the specialists can penetrate and pass through the organized remote bioenergy channel as if with a laser beam along the centrifugal and exodic paths, to monitor the cerebral nerves and to receive sensitive information for their structure, to register and assess the occurred changes. During the passage along the vagus the diagnostic bioenergy channel detects the vibrations of the pain and it is led out of the body - after remote palpation of the core of the nerve. The medical specialists could monitor the causes for headache only in several minutes. Because they are going to receive precise sensitive information about the place of its formation, about the substance, they will feel the inflammatory process, if any.

Here I will share also one more experiment which could be used from doctors - specialists for receiving of information from the central nervous system, the spinal cord and most frequently from the twelve cerebral nerves.

A picture of the head or of the spinal cord is used, made with magnetic tomographer. It is better the patient to be present, but the objects of the experiment are the already made pictures. The pictures are placed on a wooden rack, vertically, in order to be in front of the eyes for the carried out method. Then the left hand of the operator (let's call so the one performing the procedure) is turned to him in the region of the solar plexus, and the right – directed by the concentrated thought to the place of monitoring. A remote bioenergy channel is organized - beam with which the specialist passes through every problematic place of the magnetic tomography and feels the vibrations which it emits. When one knows also to what pathologic process correspond the found vibrations and thermal signals (which was related in the beginning of this part) more precise medical conclusions could be drawn. In this way in some diseases could be continued also with therapeutic procedures from the pictures with the inevitable participation of the patient. In such a way precisely are monitored also the injured parts by brain hemorrhage. After this the therapy could be continued along the same way, remotely, already in the absence of the patient. But this is another point...

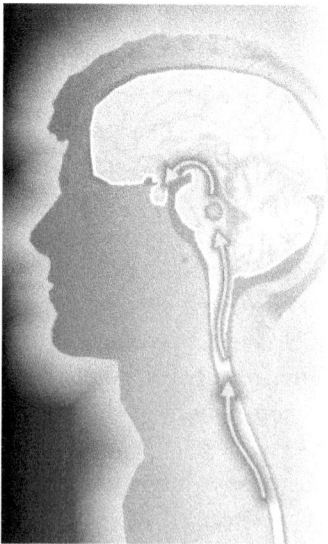

Fig. 22. I concentrate in the vegetative plexi of the abdomen, I pass upward along the pathway of the vagus to the esophagus, and I climb upward along the carotid artery and enter its cores, situated in the medulla oblongata

In cases of the diagnostics with organizing of remote bioenergy channel from nuclear magnetic picture of the patient comes out the same luminous visual picture, like in the diagnostics from nature. In projecting it on the white wall after a second of waiting appear the details which correspond to the intersected vibrations in the fingers.

Diagnostics of the spinal cord

The patient is lying on his back on the medical couch with hands, resting aside his body.

I program myself and organize the remote bioenergy channel. In the beginning I pass along the cerebro-spinal liquor from the bottom to the top to the head. On my fingers sticks information for the substance of the cerebro-spinal liquor and its quantity. In the presence of hernias high frequency vibrations are registered in my fingers. If the process is activated a warmth is also felt. If the process is an old one a cooler signal is felt.

The second passage again from the bottom to the top is along the grey matter. Due to its structure and functions the energy in the information channel is with higher frequency and flows like direct current with uninterrupted signal. If there are problems the energy along the channel stops to flow and the channel is broken. In most cases the information bioenergy signal determines the direction and leads me to the problem. Depending on the problems of the patient along the grey matter after programming of the thought, one could pass in fore columella, where are situated the exodic neurons from above - downward and after this - in the hind columella, but from below-upwards, in order to receive an information for the centrifugal paths. In parts where paired nerves come out most- frequently the signal from the cervical, thoracic and sacral segments is broken. The vibrations normalize after the therapy.

In the same manner is passed also through the white brain matter. Here however, my left hand is at a distance of about ten centimeters from the tempal part of the head, where the motory cortex of the brain is situated. With the right hand I carry out the monitoring of the spinal cord and my left hand during all the time fixes the energy field and gives me the opportunity to feel the signals from the information channel, which are formed during the passage along the spinal column with my right hand.

I pass first along the posterior tract of the white brain matter from below upwards, because here are placed the centrifugal fibers of the deep sensitivity and I continue to their cores in the medulla oblongata. If everything is OK at the end of the pathway I have to feel in the fingers of my left hand the vibrations of the cores. This remote bioenergy channel has medium frequency vibrations and medium acceleration and cooler force. When I pass along it, I always feel some type of dull and fine tingle. It increases in patients who had been cured from the Mennier's syndrome.

I pass along the side track of the white brain matter when the patient has problems with the co-ordination of the movements. And from the cervical sectors the problem frequently reflects also in the sight. In order to pass along the centrifugal pathways I use the remote channel from below - upwards. For the exodic pathways which lead its beginning from the motory zone of the cortex I use my left hand for fixing. In full concentration I pass downwards through the side pyramidal tract, through the red core of the midbrain and from there I penetrate in the side tract of the white brain matter.

In the passage along the fore tract I also use my left hand for fixing the energy above the vertex. With the right I concentrate and enter in the fore tract of the white brain matter in the cervical region and continue downwards. In the region of the thoracic vertebrae the information channel increases its frequency and in my left hand I feel the vibrations from the cores and from the motory zone of the cortex, which repel distinctly my fingers. I move my left hand slightly to the left, then to the right in order to feel the vibrations of the hectic by the information channel cores. Until now I have always fixed eight cores during my passage along this pathway.

The diagnostics of the spinal cord could be carried out also with the help of nuclear magnetic or X-ray tomography in the presence of the patient. I want to explain why is necessary the presence of the patient. Upon the organizing of the information bioenergy channel and the concentration if the patient is at the place where the diagnostics and healing have been carried out he will be stimulated with energy. This will have a favorable effect on his overall condition, will be "blown", as I use to say, his essential energy and anatomic tracts. If he is away he will be deprived of energy. An exception could be made only if the patient is in hospital and in the same building.

I had one case when I made such remote diagnostics from a photo of one climber. Day before leaving for the world championship he had fallen from the highest point of the training wall. His mother stood by me scared and tried to induce me to make the diagnostics at her risk, because at night in the hospital he had been inadequate, didn't talk and his right arm was broken. I started the diagnostics with the penetration in his cerebellum. I cleaned quickly all parts in order to receive equal medium frequency vibrations. I passed along the spinal cord and found out that there are no irremediable injuries. Only I felt that also his other arm was broken in the wrist

Together with the diagnostics I made also remote stimulation with energy of the injured places and of the whole body of the sportsman. When his mother went back to the hospital his condition had improved significantly.

The visual picture in diagnosing of the spinal cord has bright and bright pink light around the core corresponding to the grey brain matter. The light in the cerebrospinal liquor is light orange but also bright. Upon projecting of the visual picture on the wall in the moment of the passage with the remote channel start twinkling luminous effects with radiant structure. After closing of the eyes and waiting about one-two seconds, the image is projected and again the details are traced if there is need for that. In the presence of more severe diseases of the injured by the pathogenic process places is observed color and energy deviation from the described herein above characteristics.

Measuring of the waves of the brain by the sensitivity of the hands

For the first time we made such experiment in the Laboratory of Bioenergy. I had to measure the brain waves of the patients and to determine their frequency with my hands and immediately after this assistant professor Vera Tocheva made electroencephalographic monitoring. The intercepted waves by me and by the electroencephalograph coincided. We measured the brain waves of people suffering of different nerve diseases. The results from these tests I use even now when had to supplement or affirm some of my diagnoses.

In this way I can determine also the extent of concentration in executing one or other activity. The visual picture and the vibrations of the brain waves give sufficient information in order to determine the extent of concentration and the reserves in this respect.

How do I perform the method?

The patient is seated comfortably on a chair with his back towards me. I concentrate in the brain cortex at the level of the first cervical vertebra. My hands are with inverted palms. After the passage upward about five centimeters, I divide them and start to make circular movements with small diameter. My right hand works above the right half of the head of the patient, and the left - above the left half. Depending on the frequency of the received waves the length of the information channel is also determined. In the beta waves it could reach to one meter and in alpha waves to 30-50 centimeters. After this I face the patient. I concentrate myself in the middle of his forehead I organize the information channel and in the related already way I make the movements with both hands in order to receive an information from the brain cortex.

The brain waves are a type of electric energy produced by the brain. They have received their names from the Greek alphabet they are divided in four types: beta, alpha, tita and delta.

The beta waves measured with EEG are with frequency 14 and more Hz (cycles per second). The received vibrations in the tips of my fingers above the head of the patient are high frequency and register quick and strong pricks. The color which I see about the head passes from dark ink-blue to purple-violet if the activity of the man corresponds to its concentration. Every mental change I see like sparkles of which I hear the clicking as well. Or if a man works concentrated and creatively described herein above picture appears. If the two hemispheres the left and the right have equal frequency of the vibrations then also all pathogenic processes are excluded. If in some of the places above which I have placed my fingers I notice even the slightest deviation I direct my attention to it. I monitor it. Such deviations are observed in logo neurosis, in cases of epilepsy, in cases of localization of inflammation processes, which the information channel precisely determines and fixes.

The alpha waves are more clearly expressed in state of rest and are measured with EEG with frequency from 7 to 14 Hz. The received vibrations in my fingers are with medium frequency, fine and the color which accompanies them are from blue to pink. This state of the waves is registered before falling asleep, after waking up, in the starting phase of meditation, in relaxation and other similar states. In cases of hyper attention and fright neurosis upon performing the techniques for passage in alpha level occur vertical irregular waves with high frequency.

Tita-waves are with frequency from 4 to 7 Hz. The information channel fills with fine, low frequency vibrations and the color is light violet to pink. In such a state of deep sleep fall children who suffer from nocturnal enuresis. The monitoring of this state helped me to use it in my practice for treatment of this disease.

The delta-waves are registered during childbirth and death of a man and I haven't had the opportunity to monitor them sufficiently in my practice. I have noticed how before the exitus lethalis the signals in the information channel almost disappear. The light of life, as I call it fades away and gradually is lost sometimes two hours before the exitus. During these hours before the exitus the light of the mental body gets detached to a meter above the head and shines with golden light.

Where is situated "the light of life"? It is inside us. In order to see it we must bend slightly our head and place our thumbs on both sides under the eyebrows inside to the eyeball. Then in the side corners of our eyes comes out dark blue circle with a bright yellow light around it.

When we are concentrated and perform some mental activity our brain waves are respectively in beta-level. If then we look at "the light of life" we shall see that the blue circle is bigger, very dark blue and the yellow band around it is wider and brighter.

In state of rest and in a moment of relaxation the human brain emits alpha-waves. Then this light has clear beautiful colors – sea blue and sunny yellow.

Depending on the level of the brain waves the light of the circle and its size change. I have tested children with undoubted epilepsy who every day observed and wrote down the colors of this light. One of these children, Donika, for twenty days just once had seen the beautiful blue light. In most cases she told me that the light is as if powered with water, diluted.

This is an aged method for diagnostics which if studied sufficiently and mastered, could reveal to us the capacities for early diagnostics of many diseases.

Sensations in some diseases of the nervous system

Neuritis. Most frequently it is a result of cold, exposure to draft and moisture. The organized remote bioenergy channel directs my right hand to the localization of the process. I feel the respective nerve like a stretched rubber string from which I receive here and there low frequency pricks. Depending on the time when the pathogenic process had started and whether measures for his curing have been taken, I feel also repelling vibration waves. From practical experience I know that this disease is difficult to be treated with massage.

The luminous picture is muddy violet light above the localization of the process.

Neuralgia. The information channel is filled with circular motive forces. It increases its speed and my hand passes quickly as if in vacuum through a space of disseminated pain. The energy in the respective energy channels in the proximity of the injured places is disorganized. One of the most important actions in the diagnosis is to restore them in order to help the man. Most frequently this happens when I pass remotely along the meridian of the triple heater. I have a lot of relatives who had such girdle pains also from herpes zoster. Most frequently the information channel leads me to the beginning of the pain, in the spinal column from where originate the medullar nerves.

The visual picture is grey mist.

Plexitis. Radiculitis. In cases of plexitis in the region of the cervical vertebrae I feel a disintegration of the remote bioenergy channel and increase of the electric charge. In cases of passage downward along nervous pathway to the hands the channel is restored as the vibrations which I receive from the nerve of the injured arm are like from a stippled line, interrupted, with medium frequency, precisely determined interval and specific, different from the others pricks. I can determine it like electric prick. During my passage above the blades I feel again cold pricks with electric charge. The stronger the pain of the patient the colder is my sensations. In cases of acute form of plexitis these sensations pass downward from the place of injury along the lumbal and sacral section of the spinal column.

The visual picture from the injured parts resembles this of a TV monitor when we cannot establish clear connection from the transmitter and we say that on the monitor we have snowflakes.

In the presence of *radiculitis* the pain is localized and the diagnostic bioenergy channel attracts my hand to the place of inflammation between the vertebrae where the injured nerve comes out and after this leads me along the pathway of the nerve. In cases of these diseases very refined skill and experience is necessary, in order to discern correctly the sensations. Because they are variable and for a short time are received almost all types of sensations, which I have tried to classify in the beginning of the description of this method. From the place of the injured nerve luminous effects are observed with electric charge, they shine suddenly upon the passage of the organized information channel.

The paralysis stops the action of the information bioenergy channel and I do not receive information from the injured place. If anyway in the presence of paralysis there is opportunity to organize a channel and to receive even fragmentary vibrations there is a chance for healing.

Brain stroke (apoplexy). Through the information remote channel I can determine precisely the size of the impaired place in the brain. It is also a basic means for influencing? in curing. The vibrations which I receive from there are usually low frequency pricks. They depend on the type of impairment - ischemia or hemorrhage. The diagnostics itself helps for the insufflations, cleaning of the tension and for normalizing of the vibrations in the injured centers. In such a way could be made also prognostic diagnostics of people who are endangered by brain stroke. The diagnostic bioenergy channel receives the vibrations of strong pricks and starts to move with line of a lightning as the sensation for electricity in my fingers grows.

Disseminated sclerosis. Through the information channel I detect the parts from which I receive single repelling effect; the sensation is like fine tapping. The visual picture of this disease is like grey crown around the head. There is no graphite outline of the electric counterpart. If the disease is in starting stage the outline is stippled.

Diagnostics could be made also from an x-ray picture.

Brain tumor. In contact with the remote bioenergy channel the tumor organizes a core with fine repelling vibrations which change quickly its place. I have the feeling that it is a living being, which is hiding and coming out all of a sudden. At the time of execution of this method I observe a light grey sphere with a diameter from 1 to 5 centimeters to move around the head of the patient. In the centre of the sphere is organized strong and bright light which pulsates like a speck. From the outside the sphere has no outline and could increase or decrease its size during its movement.

Nocturnal enuresis. The organized remote bioenergy channel finds at bodily level precisely the causes for the disease: spina biffida (non obliteration of vertebrae in the spinal column), inborn anomalies in the excretory system. In all cases when I have met with this problem, the psyche of the children had been traumatized as well. Simultaneously with this some of the patients have had deformations of the sacral vertebrae. Changes which on their part sever the link along the centrifugal pathways of the bladder to the central nervous system. Obligatory I perform the diagnostic in the described way of the spinal cord and the cerebellum. Usually such children sleep very deeply and when their bladder is filled they do not realize the signals received from the viscereceptors of the centrifugal pathways.

In patients with this disease I register also deviations in the alpha and beta waves. It is also found out during monitoring with an encephalographer. I have had two cases with inborn anomalies: in the first the child was with double kidney basin and in the second – with a double renal duct. In their monitoring the diagnostic bioenergy channel was transformed in two parallel channels with algor characteristics. The electric counterpart shined with orange color.

Migraine. The causes for headaches are many and complex. With the help of the remote bioenergy channel always at first I detect the place where the pain is concentrated. After this again through it I feel motive forces which aim to lead it away to the outer end of the brow, to the eye or the ear.

In the presence of migraine is organized also an electric field with greater span outside the outline of the head. In this field I feel vibrating electric pricks below medium frequency.

The visual picture is like one grey field in which twinkle microscopic particles with electric charge. After leading away the pain through the known extracting of the energy in the channel the visual picture becomes lighter and respectively - its electric charge diminishes.

Neurosis. General characteristic.

In all types of neurosis the diagnostic bioenergy channel receives acceleration to the central nervous system that is to the head of the patient. There are created high frequency vibrations with repelling waves which remove my hand from 50 to 100 centimeters depending on the stage of disease. It is necessary to clear the tension with the already known to you procedure for extracting. Great tension appears also in the passage along the energy channels. The tension there should also be cleared in order to create conditions for precise diagnostics.

The visual picture of these diseases appears almost immediately but it frequently changes. That is why along with the supply of energy one should carefully trace the oncoming changes until the picture is clarified after the tension is normalized.

I monitor the spinal cord and especially its posterior and anterior columella, respectively — the centrifugal and exodic pathways. I feel one specific tension filled with electric charge which at precisely specific intervals sends equal in force and duration pricking signals. I have the feeling that I pass along a stipple line. Inherent to this state is also the tinnitus which is heard. It comes from the level of the medulla oblongata and continues downwards along the exodic pathways of the spinal column. Upon repetition of the procedure several times the tinnitus fades away.

The fear neurosis in the cases which I have met with always has been accompanied by changes in the region of the cervical vertebrae and vestibular apparatus. During the diagnostics I receive visual images of life situations that had caused or intensified the disease.

In the presence of *logo neurosis (stammering)* I form a remote bioenergy channel which leads me above the centers of Brocca, above which is organized a whirl with the form of a funnel. From there flows out energy, the power of which is connected with the stage of the disease. But in all cases pierces the graphite line of the electric counterpart. During my passage downward – to the facial part of the head I perceive single coarse oscillations depending on the spasms of the muscles in this region. The treatment of this disease is complex and in it should participate all, with who the patient is in contact.

Epilepsy. It happened so, that already in the first years of my practice I met with children who had this diagnosis. Characteristic for it is the sensation for localized cold regions with different size. The information channel accepts their energy form, increases its power fields and is filled with helical repelling vibrations.

In the visual picture above the registered zones I notice dragging downwards the graphite line of the electric counterpart. A luminous funnel with its top upward comes out of the place where the process was localized.

In *schizophrenia* I feel the injured focus more-inside and deeper. With every touch with the remote bioenergy channel it wants as if to hide its size. The sensation for this comes from a spiral whirl with repelling force. The information channel is filled with this energy and the distance for making the diagnostics is increased reaching to above one meter. In the visual picture in most cases around the head graphite outline of the electric counterpart is broken. This means that the thoughts of such people very quickly come out of the orbit of the biofield and spread in space or reach with higher than the ordinary velocity to mentally selected object. The vibrations of the cerebral beta-waves surpass 25 Hz but actually are high voltage waves demonstrating the localization of the process.

As soon as the information channel enters in their field a movement starts similar to the movement, which marks the hand of the encephalographer. In such patients if one enters in the hypothalamus cool and harsh pricks are felt coming from the stored in greater amount serotonin.

Diagnostics of sensory organs
The Eye

One cannot contradict to the old saying that the eye is the mirror of the soul.

Because we really feel how it, the soul, vibrates in every particle of the eye. Our eyes have the capacity and force not only to emit our feelings, thoughts, spiritual and bodily sensations, joys and sorrows. They radiate also pure healing energy. Thanks to the eyes and our mind we succeed to penetrate in the matter, to see so to say its density, color, vibrations.

The patient is seated on a high chair or so that his eyes and the eyes of the operator be in one horizontal plane.

The distance between them is about one meter.

I carry out the method with both hands as in all double organs. So the information is supplemented also with details from the matching of both organs. (Fig. 23)

I organize two bioenergy information channels (beams) which come out of the middle fingers of my hands. The beam from the right hand is directed to the left eye of the patient and this from the left – to his right eye.

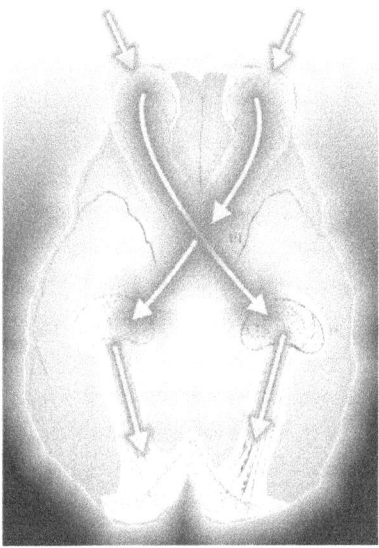

Fig. 23. I carry out the method with both hands as in all double organs. So the information is supplemented also with details from the matching of both organs. I organize two bioenergy information channels (beams) which come out of the middle fingers of my hands.
The beam from the right hand is directed to the left eye of the patient and this from the left – to his right eye

I start with circular movements clockwise simultaneously with both hands as their diameter is corresponding to the size of the eye ball.
I direct my thought consecutively to:
1. The outer stroma membrane of the eye ball. I receive information about its soundness and size. In normal state without special deviations the information channel is with equal acceleration and in the middle of my fingers I read medium frequency vibrations.
2. The medium layer consists of vascular membrane. I feel its coarse structure. If it is injured in certain places by clots I receive single pricks in my fingers and so I determine precisely their locality.
3. I pass to the front pole of the eye ball, let's call it the window. This I make again with circular movements and receive information from the cornea. I feel its bulge and substance. If it is clear an uninterrupted channel with constant frequency is

organized. If there are changes in the transparency or any disease the vibrations are changed.

4. I make helical movements from the pupil to the end of the iris. The energy in the remote beam increases and I feel high frequency pricks from the whole surface of the iris. In such a way I receive information in pads of my middle fingers for every microscopic particle of it. As I take into account also my knowledge of iris diagnostics, the vibrations give me opportunity to find out also some details or confirmations, received from the previously diagnose organs of the patients. Here always comes out the higher tension in the beam-channel and this hinders the diagnostic process. I have to balance the tension – I pull it back towards me and grind it with force in my palms. I repeat this several times - until I am sure that I can go on.

The iris is a projection of the whole sympathetic nervous system. Anatomically the sympathetic nerves from the lower part of the body lead to the lower part of the iris, and from the upper part of the body - to the upper part of the iris. In the performance of the helical movements my fingers record the precise place of the deformations. If I concentrate my look I always will detect white or black spots, which have with circular shape or appear as beams, arches and transversal lines.

The pigmentation of the iris is individual and is determined by the inherited characteristics and racial marks, apart from this also by deviations, defects and other influences on the organism. By the pigmentation one can judge for the opportunity for occurrence of some diseases. For instance there are people whose iris is surrounded by a white band which could be noticed with unarmed eye. For such people are inherent diseases of the circulatory system, anemia, and liver.

5. I organize with my thought, with my hands and my look a bioenergy beam and fix the pupil. Immediately I register its size and form. I compare the differences between the two pupils. The changes in the width of the pupil are realized by a refectory mechanism, by the receptors situated in the retina. During the diagnostics they help to increase the power and passage of the information energy channel to the retina and the visual nerve.

I concentrate myself and penetrate inside with the bioenergy channel-beam. I pass through the pupil, the lens, I penetrate in the retina. I feel a humid substance between the cornea and the lens. In the tips of my middle fingers a circle is as if stuck with the size of the lens and I feel its jelly-like substance. The sensation comes from about 10 millimeters inside I discern the vibrations of the medium which pulsates, shrinks and expands. Even with the smallest problem in the lens the bioenergy channel is blocked and must be organized again - through entering for a second time through the pupil.

I go further inside the space between the lens and the retina. I feel a jelly-like protein substance of the vitreous body. If the pressure in the anterior and posterior eye chamber is higher than the normal during the passage through it (I have in mind the vitreous body) vibrations with attracting and repelling effect are detected sometimes. I balance the tension in the channel with the help of the familiar technique for extracting of the energy in the channel in reverse direction - towards me and so I decrease the tension.

I enter again through the pupil and reach to the retina, through the blind spot I go out from the eye ball and continue along the ophthalmic nerves of both eyes to the place where they cross. This place is known under the name optic chiasma. Here in the bioenergy channel the energy is as if attracts my fingers and they almost join.

In the presence of myopia these forces attract my fingers slightly in front of the normal place of the optic chiasma. The bigger the myopia the more forward I receive the sensation for the energy attraction of the chiasm. In such cases I have the feeling that the ophthalmic nerves are shorter. From experience I can determine also the diopter.

In hypermetropoia the energy channel-beams coming out from my both middle fingers are attracted slightly backwards than the normal place of the chiasm. I feel where precisely it is and where the two beams should cross but this does not happen there. At the same time I feel also one extraction and attraction backwards. As if the ophthalmic nerves are longer. By the force and the length of extraction I can determine the diopter.

Sometimes I feel also different tension of the bioenergy channel in the two eyes. While nearing the place of the chiasm one beam becomes shorter than the other. The forces of attraction are irregular and I feel distortion of the chiasm in the one end downwards. This happens when the two eyes have with different diopters. Then I realized the treatment by attempting to balance the attracting forces in one horizontal line. Thus I penetrate in one process in which depending on the causes and the duration of the disease, I can restore the balance in the chiasma and from there to diminish the diopters or to bring them to values closer to the normal.

After I diagnose the chiasma I follow the pathway of the bioenergy channel which crosses my hands and they come out in the occipital part. My right hand comes out in the right occipital part of the patient, and my left hand - in his left occipital part. Every problem in this tract after the chiasma is also registered. Interruptions, tension and increase of frequency of the vibrations are felt – according to the problems.

The visual scheme which I project in every stage of the research is different. The head of the patient shines with a yellow light. If the iris is brown or black the eyes are outlined with an orange light while during the energy penetration inwards - this light becomes more intense. If the iris is bright - blue, green or grey the eye ball shines with pink-violet light which also becomes more intense with my penetration inwards. Sometimes I see the two eye balls to shine in different hues, one to be brighter than the other. Then the problems are in the eye with darker hues. Otherwise said the deviations from the normal color with darker spots or with darker pink reddish hues the inside of the eye balls. After coming out of the energy beam from the occipital part of the patient I see on the white wall above him like on a screen a luminous scheme of the pathway along which I have passed.

At the beginning of implementation of this method one can use an anatomic scheme for the sequence of the procedures positioned on the wall behind the patient.

Sensations in some ophthalmic diseases

Glaucoma. The increased internal tension of the eye has repelling vibrations and severs the directed by me energy beam already in the very beginning. After I balance the pressure with specific movements and succeed to penetrate through the pupil inside to ophthalmic nerve I receive one more proof for the presence of this disease. It is my sensation that the ophthalmic nerve is as if pushed by the eye ball. Until now all patients with proven glaucoma who have undergone through my practice have had also concomitant ailments like spondylosis, impairment of the flexibility and structure of the cervical vertebrae.

Cataract (lenticular opacity). The diagnostic bioenergy channel-beam is interrupted by cool repelling vibrations in the region of the lens.

Lazy eye. I register very weak vibrations in the information channel. The organized energy beam cannot penetrate in the eye ball and move along the nerve to the chiasma. The eye itself comes inside and with the course of time begins to be noticed. I have had several such cases with young people who had this medical diagnosis after trauma. I taught them to perform every day specific energy exercises, with which to maintain the activity of the eye and stop the process of rudimentation.

Due to the organized bioenergy remote channel-beam one could already during the diagnostics achieve dissipation of the blood clots in the eye. I've had several such cases.

The Ear

The diagnostics is performed from a distance of about 50 centimeters.

The patient is seated with his back towards me on an ordinary chair.
I organize two remote energy channels with both my hands. My right hand monitors the right ear of the patient, while my left hand - the left ear. (Fig. 24)

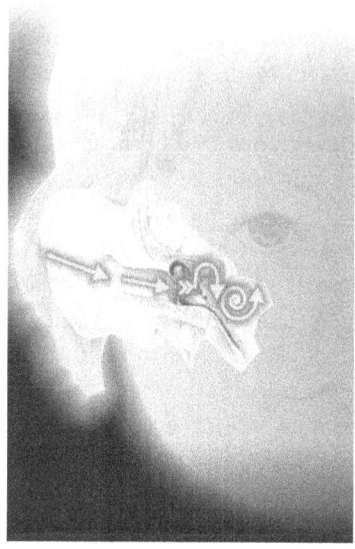

Fig. 24. I pass along the pathway of the sound waves; I excite the receptors with the energy beam and receive information in my fingers from every particle along this pathway

I pass along the pathway of the sound waves; I excite the receptors with the energy beam and receive information in my fingers from every particle along this pathway.

I penetrate through the external auricular channel to the tympana membrane which separates the external from the middle ear. If the tympanic membrane is punctured in the passage is felt cold. In an inflammatory process in the ear I feel repelling vibrations which break the bioenergy channel. Then I start to make circular movements near the auricle by screwing in direction from the back of the patient upward, forward and after that downward. Or actually I make a circle around the auricle with a diameter slightly bigger than it.

A field is organized in which the tension grows and gradually pushes aside the hands making a circle. The pain is extracted in such a way and the tension in the ear caused by the inflammation process is normalized. After this procedure I organize again the remote bioenergy channel-beam and penetrate through it from the tympanic membrane directly to the maleus which in normal state is inosculated firmly with it. I receive information about its size (normally about 7 millimeters) and from its head I pass to the incus. I define my feeling for the substance in this part as fine crystals. I accept the incus as a crystal bridge. The channel-beam curves at one end of the bridge where the stirrup (stapes) is situated and directs itself to "the fenestra ovalis" of the internal ear. Provided everything is in the normal limits the information beam is uninterrupted and with normal frequency. All deviations from the normal state vary the tension in the channel. Attracting and repelling vibrations with different frequency are received accordingly.

I pass along the three semicircular channels: lateral, posterior and upper and I pass inside the cochlea by screwing inside two and half times. After that the channel-beam continues respectively along both nerves whose cores are in the Pons Varolii. Here both channels cross and my hands respectively come out in the opposite audio zone of each ear. Above the cores of the nerves in the Pons Varolii I feel high frequency prick in the fingers.

If the information bioenergy channel interrupts in the inner ear I go back and penetrate again, always from the exterior otic channel.

When the exact anatomic pathway is well known and the organized motive forces in the information channel are followed, in the fingers of the diagnostician will be reproduced sensations for all deviations from the normal state.

I receive a visual picture upon penetration in the middle ear. To the penetration along the nerves in the Pons Varolii I see a scheme of the nerves and their cores. In a normal state the color of the scheme is golden yellow.

Sensations in some functional disorders and diseases of the ear

Inflammation of the middle year. In the presence of chronic processes I feel a humid cold substance. In the presence of acute inflammation processes due to tonsillitis and other infections - hot repelling high frequency vibrations.

Catarrh of the duct. It occurs as a result of sinusitis, nasal cold, enlarged tonsils or swelling of the mucous membrane of the throat. An obstruction of the auricular duct occurs which leads to pains, reduced hearing, feeling for encumbrance and noises. My sensations represent warm repelling vibrations which dissolve the forces in the information bioenergy channel.

Reduced hearing, as a result of noise influence. It is formed as an occupational disease. I have special observations with weavers. After the fifth year of work in the mill tinnitus starts inside their ears, a misbalance of the intraear pressure is observed. During monitoring I feel interruption of the information channel along the pathway to Pons Varolii. I have the feeling that the nerve is broken from its core.

Otosclerosis. In the presence of this disease I feel in the region of the hammer (maleus), the anvil (incus) and the stirrup (stapes) an ossification, a cementation. I can't separate their vibrations separately, as it happens in a normal state.

The Nose

Two years ago patient of mine boasted that when he went home after the therapy, he felt the smell of the meal cocked by his wife. This happened to him for the first time in the last twenty years. And this had happened since I haven't paid special attention of his disease as we were treating another disease. This concomitant effect was the result of the fact, that during each therapy I pay attention to all organs and systems and work for the harmonization of their functions.

The diagnostics here as well as in other cases is also like a therapeutic method.

The patient is seated on a chair with his face towards me.

I direct both my hands with almost clinging palms; I concentrate my look and thoughts in the basis of the upper part of the nose. In such a way are organized both information channels. Gradually I separate my hands and they pass on both sides of the nose. Before its thicker part they bend to one side along the lower rim of the zygomatic bone. I continue upwards and I come out to the sensory zone of the ears, after which I turn along the outlines of the brows and come to starting position. Upon repetition of these movements several times appears also the visual scheme which together with the sensations in my fingers gives me an opportunity to specify the exact diagnosis of the obstructed sections, to see eventual fractures of the bone and other changes.

For the elimination of these problems in my practice I had created techniques for self treatment which give very good results. After recurrent operation of polyps in her nose I taught a patient of mine to treat herself and already five years there is no need to undergo an operation.

Cardio vascular system

The first procedure with which I start each diagnostics of the cardio-vascular system is the check of the blood pressure of the patient. I do it in the following manner:

I stand at a distance of 50 centimeters from the patient. He is standing in front of me with freely relaxed hands and with palms turned towards me. I organize an information bioenergy channel with the thought to detect the vascular pulse in the region of the wrist.

According to the characteristic of the waves and the frequency of the vibrations of the pulse I can classify the received information in three groups:

First group – high frequency vibrations with distinct repelling wave. They are characteristic for high blood pressure with systolic limit above 160 mm mercury column or diastolic - above 100. Pulse waves - above 80 per minute.

Second group – medium frequency vibrations with fine distinct wave. They are characteristic for normal blood pressure in the limits of 120-80 mm mercury column. Pulse waves - 65-70 per minute.

Third group – low frequency cool, difficult to discern vibrations. I intercept them when the blood pressure is low, in the limits of 90-60 mm mercury column. The pulse waves are below 60 per minute.

Analyzing the received information I can determine the blood pressure in the moment with precision to 10 mm.

I drag upwards along the pathway of the artery (ulnaris) the already organized remote bioenergy channel and find the next push of the pulse, which fixes the limit of the blood pressure at the moment. It will appear at a distance of 60 to 100 mm mercury column from the first pulse in the region of the wrist. So I receive the exact value of the diastolic limit.

I continue upwards along the same straight line and in the tips of my fingers I feel one more pulse. With it I read the exact limit of the systolic blood pressure. In the cases when the blood pressure of the patient in the moment of measuring is unstable the diagnostic channel vibrates in limits which I can determine precisely. Such are also the cases when the blood pressure is high. I feel it already in the beginning and together with the techniques for distribution of the energy I apply also techniques for normalizing and stabilizing of the blood pressure. Here are two of these techniques:

1. The patient is seated on a chair. I put my right hand with the palm on his forehead and with the thumb and the forefinger of the left I press in the base of the occipital area. After three to five minutes the blood pressure starts to drop evenly.
2. I pass remotely several times along the energy channel of the heart until I balance the tension. I stop when I receive uninterrupted signal with medium frequency vibrations.

In all types of tests the measuring of the blood pressure has important diagnostic significance for all organs and systems in the human organism. The measuring in the described manner is accurate, without pain and free from stress.

In my practice I have reached to the conclusion that the blood pressure in every man has specific limits within which he is feelings comfortably and this must be kept in mind in the carrying out of every treatment.

Dangerous are the cases when some people do not feel any signs whatsoever for rising of their blood pressure but during measuring it is with high values.

From my practice I can say also that most people with inherited problems in the gastro-intestinal tract have low blood pressure up to the age of 30-35. After this age it starts to rise.

The diagnostics of the cardio-vascular system is carried out along the pathway of the systemic and pulmonary blood circulation.

The more detailed this system is known, the more information could be derived during the performance of the method.

The passage with the remote bioenergy information channel along the systemic and pulmonary blood circulation is one, so to say, remote bioenergy massage of the blood vessels and the heart. It stimulates their activity, raises their elasticity, and cleans their walls from depositions. It contributes also for the cleaning of the path of the blood flow and for good and quick irrigation of all parts of the body. This plays role also in the therapy itself. It influences favorably the elimination also of other problems which are not linked with circulatory system. (Fig .25)

I organize a remote bioenergy channel at a distance of up to 50 centimeters with starting point in the hilus and I move my right hand clockwise in order to receive information about the size and the position of the heart. Sometimes there are deviations from the normal position of the heart. It could be situated slightly to the right or exactly in the middle of the chest, enlarged or diminished. In my work with sportsmen I traced how the muscle masses of the heart increase, but this never made me happy. Because in such cases a special tension is felt already in the first remote touch. Parallel to this in the channel are created motive forces which lead my hand downward to the sixth intercostality. This is a signal recurring always in cases of cardiomegalia.

After I receive information about the size and the position of the heart I concentrate and penetrate inside to the three different layers of the myocardium.

The wall of the heart is built up of three layers: inner—endocardium, middle—myocardium and outer—epicardium.

The thickest and of greatest functional importance is the *myocardium*. It is built up of muscles of specific type, which contract rhythmically and independently on our will. The thickness of the muscle layer of the atriums is 2-3 millimeters, of the right camera - from 5 to 8 millimeters, of the left camera - 10-15 millimeters.

The endocardium is a thin membrane, which covers from inside the myocardium and through duplicators forms the vellums of the two athrio-ventricular valves.

The epicardium is a smooth, thin and transparent membrane, which eases the friction between the heart and the walls of the pericardial sack. It is its outer layer.

When penetrating to the pericardial sack I detect cool low frequency vibrations and a sliding substance, which corresponds to the quantity of fatty tissue, stuck on it. It depends on the quantity of body fat and encumbers the heart, if it is in proportions above the normal. Then we speak about fatty heart.

I have had also cases, when I detect the already proven from medical tests calcification of outflow in the pericardial sack. My sensations are like the ones when I try to penetrate in cemented space. The vibrations are similar to those in calcification of the discs of the spinal column but less repelling.

I penetrate in the middle layer – the actual muscle of the heart – *the myocardium*. I palpate the right side of the atrium, which is thick between 2 and 4 millimeters and pushes the blood only along the pulmonary path to the lungs. Here in the remote bioenergy channel are created helical motive forces in upward direction. In a given moment I start to follow them. This moment coincides with the passage of the venous blood from the right atrium into the right camera and its expulsion in the pulmonary artery from where starts the pulmonary blood circulation.

I follow it by including also my left hand at the place of forking of the artery respectively to the left and the right part of the lungs. Two remote bioenergy channels are immediately formed, which follow the blood in the thin capillary net around the alveoli. Here, through my fingers, I register change in the vibrations of oxidation of the hemoglobin of the red blood cells with the oxygen from the air in the lung bubbles. When penetrating in the pulmonary artery I have the sensation that something like magnet attracts my fingers along the path of the arteries to the two halves of the lungs. After I register the oxidation in the lung bubbles the information channel receives acceleration, which repels my hand from the patient. The vibrations are fine, medium frequency and warm. A magnetic gravitation appears in both information channels, which fork to two more and move respectively with my middle fingers and forefingers of my both hands to the left atrium.

So with the four channels, corresponding to the lung veins, I penetrate in the left atrium. They merge in one remote channel and the information from it is registered by the tips of the fingers of my right hand.

I start my inspection of *the systemic blood circulation*. (Fig. 25) The diagnostic bioenergy channel receives acceleration in the moment of expulsion of the blood from the left camera upwards to the aorta. It makes a curve there where the coronary arteries separate, carrying the clearest blood in the myocardium. The created acceleration, coinciding with the transportation of the blood, accelerates the vibrations and the force of the energy in the channel is increased. Together with this I start to I receive even more information: I receive the high frequency vibrations of the high blood pressure, I feel the volume of the aorta. The vibrations in my fingers register even the smallest changes in the chemical composition of the blood.

In the region above the pelvis I include also my left hand because the aorta forks in two for both legs and again two remote bioenergy channels are organized. I pass through the legs and extract the tension outside and to myself. I grind the residual energy from the channel in my palms and process mentally the received information

From my practice up to now I can make a conclusion that the richer is the blood in vitamins, minerals and all other necessary for a man substances the stronger, uninterrupted and with equal acceleration is the diagnostic bioenergy channel. In such a case the diagnostics could be performed from a greater distance - more than 50 centimeters. And vice versa - the poorer is the blood in vitamins, salts, mineral substances the weaker, interrupting and with attracting force is the diagnostic channel. In such cases the distance between me and the patient could be reduced to 10-20 centimeters.

According to the particularities of the disease I register also deviations from the normal composition of the blood. For example in cases of higher body temperature I feel the substance of the blood thinner. Thicker and glutinous I feel it in cases of malignant diseases in final stage.

Fig. 25. I organize a remote bioenergy channel at a distance of up to 50 centimeters with starting point in the hilus and I move my right hand clockwise in order to receive information about the size and the position of the heart

In the presence of anemia the information channel has low acceleration, the signal is cooler, and the substance of the blood is thinner and in more severe cases becomes even glutinous

I do not pass along the venous pathway from both legs upward because the forces inside the remote channel can move upward a thrombus, if any. Moreover if during the passage downward I have intercepted single, warm, pricky signal for varications.

The following, so to say, scheme, along which I pass is a repetition of the pathway to the aortic arch and passage along the arteries of both hands of the patient, respectively with two information channels, which percept the thickness of the arteries. I receive information about the soundness and the depositions, if any, on the blood vessels, the activity of the heart; I feel also the pulse waves and the blood pressure.

In order to receive more information about the activity of the heart I penetrate in the coronary arteries. The starting position for organizing of the remote channel is the one described herein above for penetration in the left camera. After this I penetrate in the aorta and from there exactly above the aortic valve I pass in the coronary arteries. In case of oxygen insufficiency here I receive prickly dolorous sensation. If the blood vessels of the patient are pathologically changed the diagnostic channel decreases the acceleration, receives their thickness and vibrates with low frequency pricks. This is characteristic for the so called ischemic disease of the myocardium.

I also receive information during the organizing of the remote channel to one of the main reflexogenic zones - *the carotid*. It is found on the level of the thyroid gland and is the place, where the common carotid artery divides to internal and external. Here, in the wall of the artery, are situated the baroceptors and chemoceptors and could be confirmed the registered readings of the values of the pulsating impulses and of the blood pressure. From there the bioenergy channel almost automatically reaches to the cardio vascular centre in the medulla oblongata. This centre pulsates, vibrates depending on the values of the blood pressure. If it is high - with high frequency vibrations, if is low - with low frequency and in normal state - with medium frequency vibrations. In low blood pressure an attracting effect to the centers is felt.

When applying this method I cannot differentiate the separate forms of stenocardia, myocardial infraction, mitral stenosis and other specific diseases of the heart. I discern renal hypertonia, incretory conditioned hypertony, hypertonic disease. But I am sure that if used by a doctor cardiologist, especially at places, where there is no modern medical equipment, sufficient information will be gathered to order the correct treatment.

Already in the organizing of the remote bioenergy channel I receive information about *the blood group* and *Rhesus factor* of the patient.

The long diagnostic practice gave me the opportunity to find out that some patients are very susceptible when the remote information channel is used. Their organism allows quick penetration and permits easy definition of the diagnosis. With them I carry out easily also the therapy. On the contrary, with others, no matter how many efforts I render I meet with resistance. When this happens above the respiratory system, to which I direct myself at first, motive forces are immediately created in the channel, which attract my hand downwards to the stomach. Here already I receive one extraordinary sensation for the character of the channel namely: cold vibrations, equally persistent medium specific frequency. This sensation makes me indicate the blood group of the patient: "0" with RH (-).

In blood group "0" with RH (+) I do not meet with resistance in the organizing of the channel but I feel it more feebly because it is with very fine, caressing vibrations, which at places along the pathway of the system disappear and appear again. In order to make a precise diagnostics in such cases one must apply technique for organizing and stabilizing of the energy.

In blood group "B" with RH (+) in the organized channel at the beginning the signal is cool, but after an instant is filled with vibrations above medium frequency - I can compare it to dentists drill, working at low rotations. During the diagnostics of this blood group motive forces an always formed which in a most unexpected moment could vehemently deviate my hand. If even this happens I am already sure and I can indicate the group. When blood group is "B" with RH (-) the characteristics are such as I have described herein above but with distinct vehement motive forces in the information channel.

Blood group "A" with RH (+). The information channel which I organize between me and the patient is filled with fine, warm, uninterrupted signal. If RH is (-) the signal is slightly cooler, but again uninterrupted and with equal medium frequency vibrations.

For blood group "AB" with RH (+) is characteristic an uninterrupted signal, with vibrations above the medium frequency level. The result is a strong, well organized channel, with which diagnostics could be quickly performed and treatment carried out. With RH (-) the signal is the same. At certain places however there is an interruption and I have the feeling for uneven movement.

The information from the blood group helps me to prognosticate the duration of treatment. For example "0" RH (-) is cured more difficulty and requires spending of more energy by the healer.

Upon the passage along the pathway of the systemic and pulmonary blood circulation the visual picture is in light green. Around the green light with the form of the heart is seen one band of bright yellow light – like a reflection. If the myocardium has problems, this yellow light grows dim and looks like a shadow.

Spleen and lymphatic organs

In order to make thorough diagnostics I have however to palpate remotely the spleen as well. The specific thing for it is that it is abundantly supplied with blood and reacts to many diseases with increase of its volume. It receives saturated with oxygen blood - through its own artery, which comes out directly from the aorta.

I organize the diagnostic bioenergy channel at the level of the spleen and receive information about its size. (Fig. 26) I penetrate inside through its hilus and reach the spleen tissue. The vibrations are organized in the information channel and according to the blood group of the patient. So I register even the slightest imperceptible signals for problems in the spleen. The diagnostic channel leads me to hypothalamus. The vibrations, which I feel in the passage to the hypothalamus, are with high frequency, with super electric charge. I have the feeling that the spleen has its own nervous pathway to the hypothalamus and is innervated from there. So it receives information about all chemical reactions taking place there and prepares its reaction. From the visual picture, which I receive, I judge what the strength of the immune protection is. The more neon green is it, the weaker is the immune system. If the visual picture is filled with clear white light, this is a sign of strong and stable natural immunity of the patient.

To this system belong also the tonsils. They can also be diagnosed by this method if you know well their structure and position. One must know that information is stored around them about all their pathologic problems in the past years. Every their inflammation and change injures the lymph flow and could be registered by the specific vibrations, incoming in the remote information channel during monitoring.

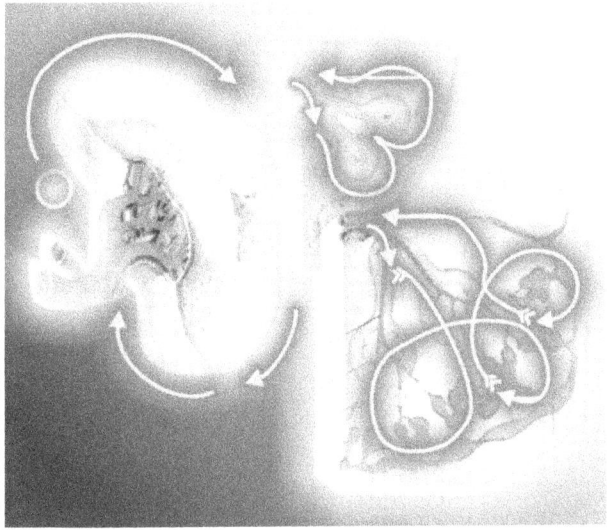

Fig. 26. I organize the diagnostic bioenergy channel at the level of the spleen and receive information about its size. I penetrate inside through its hilus and reach the spleen tissue

CHAPTER SEVEN

Diagnostics Through Remote Bioenergy Palpation Of The Fourteen Basic Energy Channels

The long development of the ancient Eastern medicine is based on the philosophic idea of the five proto elements: earth, water, fire, metal and wood. Water, earth and metal are regarded as the constructive elements of the body, wood materializes its growth and fire is considered to be a symbol of movement and progress. Air is a vehicle of the creation, it is the animating force.

Between the cosmic proto elements exist links and interdependencies and all they are dependent also on the surrounding environment, on the course of time and on the cosmic influences.

According to the traditional Eastern medicine man is a projection of the macro cosmos and is a vehicle of its basic characteristics. His body is made up of the same elements and obeys the general laws of nature. The human body is composed of anatomic physiologic units - organs, which interact among themselves and with the surrounding habitat. Each one of the organs of the body corresponds to a given proto-element.

The base of the interaction and the interdependency of the parts of the human organism according to the traditional medicine of the east is "the vital energy". It pierces the whole material world, regulates the complex relations and processes in it; it is transformed and reproduced, securing life and progress. The most essential form of the exhibits of the vital energy and its functioning is the struggle between the two opposing elements - YAN and IN.

The Eastern medicine teaches that the functions of the organs of the body - YAN have originated as a result of the use of nutritious substances - IN. In such case exists a process of increase of YAN and reduction of IN. In the same time the assimilation, the processing and metabolism of the nutritious substances in the organism - IN is not possible without drain of certain quantity energy - YAN.

In the healthy body is maintained a relative balance of these processes and it guarantees the normal functioning of the organism. When enhancement or the weakening of these two basic elements passes a certain limit this causes anomalies in the organism. And they can cause the onset of pathologic processes which the medicine qualifies as diseases. That is why the maintenance of balanced state between IN and YAN is in the basis of the vital activities of the human organism and the recuperation of the impaired balance is the most important task of the medical practice.

The idea that the energy pierces the whole cosmos and is in the foundation of life, in the course of centuries has formed in orderly philosophy and medical theory, for which development and enrichment an exceptional role has the practice of the ancient Chinese healers. Their activity is based on the belief that the energy which circulates in the human body is the same universal natural energy which gives birth to and supports life. They call it TZI.

In the human body TZI plays reproductive, nutritious and protective function. It moves along the twelve basic meridians - energy pathways and links the main internal organs with the other parts of the body and the skin. Each meridian is linked and corresponds to the activity of a specific internal organ and the circulation of the energy along the meridians is subordinated to the internal biologic clock. The complete ring of circulation of the energy TZI along the twelve meridians is carried out in twenty four hours as the maximum tension in each of the meridians lasts for two hours. Apart from these twelve energies the meridian forms a chain along which every one is linked with the preceding and following meridian.

Apart from these two more energy pathways regarded and described as meridians exist. These are the anterior middle meridian referred to also as REPRODUCTIVE and the posterior middle meridian, referred to as MANAGING meridian. We could call them main meridians as they play specific regulating role compared to the remaining twelve meridians. Most generally we could say that the anterior middle meridian governs the meridians along which flows the IN-energy and the posterior - these, along which moves the YAN energy.

Before I start the practical description of the method I want to repeat that we regard the human body as one constantly working energy system in which circulates the TZI the – universal cosmic energy. This energy flows in the body along certain channels or places with higher conductivity which is near to arteries and the nerves. Like the arteries and the nerves the channels of TZI are also covered by the body muscles. They branch and link the energy spots, supply the tissues and organs with TZI by carrying and transferring energy information. This system of energy circulation is under the control of the sensorium notwithstanding whether a man is aware of this or not.

The implementation of the method for remote bioenergy research of the energy channels (meridians) requires very profound knowledge of the energy pathways, the beginning and the end of every meridian, the most important points along its path, the direction of movement of the energy, its strength and duration, as well as the polarity of the meridians, their most active periods in the day and their critical periods in the year.

What must we know about this method?

Following the energy pathways (the meridians) in the human organism we receive most general information about the state of its organs and systems and determine the character of its health problems.

If energy circulates along all meridians this means that man is healthy.

Every deviation from the usual oscillations of the respective meridian directs us to the problems of the separate organs and systems. It should be known that to a certain extent this method is auxiliary or concomitant to the application of other already described bioenergy methods. For example, we would not be able to make diagnostics in the anatomic way of the pulmonary system, if the energy pathways of the lung meridian are blocked. In such a case it is obligatory first of all to set free the energy in the meridian, to secure its normal circulation and after that to penetrate in the system through the remote bioenergy channel for making a diagnosis.

The use of this method in the beginning of every diagnostic procedure helps for the organization and harmonization of the vital energy, prepares the patient physically and mentally. It could be also applied after the end of every diagnostic and therapeutic procedure. In this manner the healer becomes convinced that the energy in the organism of the patient flows normally, that it is balanced and cleared.

This method could be practiced also for auto diagnostics and auto therapy however by healers, who have sufficient experience.

The position of almost all meridians allows the organizing of the remote information channel. It is almost impossible this to happen for auto diagnostics along the managing meridian and this of the bladder.

How in practice is carried out the auto diagnostics. The organization of remote information channels for the meridians which are situated on the left side of the body is realized with the right hand and of these situated on the right - with the left hand. The energy flowing in the meridian from the starting to the final acupuncture spot is followed. If the meridian is in disharmony the flow of the energy along it is hindered and this is registered with the help of the information channel. Then it is necessary for me again and again to pass along the pathway of the meridian - until the normal flow of energy is recuperated. In this manner is carried out also the auto therapy.

Auto diagnostics and auto therapy along the meridians could be achieved also through visualization. With the auxiliary means of the meditation the meridian is filled gradually with clear light, which passes from the starting to the final acupuncture spot. It is very important the light to fill all blocked centers and to flow along the channel in the course of several minutes. For every meridian in the visualization is used different color light.

During the passage along the meridians the operator carries out direct stimulation of the circulation of energy and sets free the blocked centers. Every encountered obstacle along the pathway of the remote bioenergy channel - beam causes something similar to the repelling wave and these increases the distance between the hand of diagnostics and the meridian. Then it is necessary to start the already known withdrawal of the remote channel in order to set free the accumulated tension and to open the pathway of the energy. When this happens it is felt how TZI normalizes its movement from the starting to the final point of the meridian and the diagnostic bioenergy channel receives uninterrupted, vibrating with equal frequency signal from the beginning to the end.

Let us go further to describe the remote bioenergy monitoring of the energy meridians. Until mastering this technique for diagnostics, one could work by using schemes or models of the meridians.

In the pointing out of the more important standard points of the meridians is used their designation in the book "Acupuncture zones atlas" by Macheret, Lyusenyuk, Samosyuk, Publishing corporation "HIGH SCHOOL", 1986.

Meridian of the lungs

Polarity - IN. The movement of energy in it is centrifugal. It receives energy from the meridian of the liver and transmits it to the meridian of the colon. It consists of 11 points. (Fig. 27)

I concentrate myself in the starting energy point in the chest of meridian P/1/. I organize the information remote channel and through it I follow the eleven points. I start from the lung apex, I continue downwards along the internal side of the hand and I come out from the tip of the thumb.

I find out a misbalance of the energy in the acupuncture points of the meridian in diseases like: asthma, bronchitis, smoker's chronic cough, influenza, hard breathing caused by problems in the stomach and gall bladder, sore throat, different types of rhinitis, tendovaginitis and others. If the energy in the meridian does not circulate normally, a swelling resembling a small tongue is palpated in the base of the thumb.

The meridian is most active between 3 and 5 o'clock in the morning.

Most easily the misbalance could be dispersed and the normal flow of the energy restored if we use the opportunity to perform the diagnostics and therapy in the morning hours.

The normal flow of the energy in the meridian has an effect not only on the lungs, but also on the olfaction, the antrums, the skin and secretion of slime.

The position of the meridian gives opportunity to perform auto diagnostics and auto therapy - through organizing of two information channels - separately with both hands.

White, blue and pink are the colors, which influence favorably the diagnostic visualization and therapy.

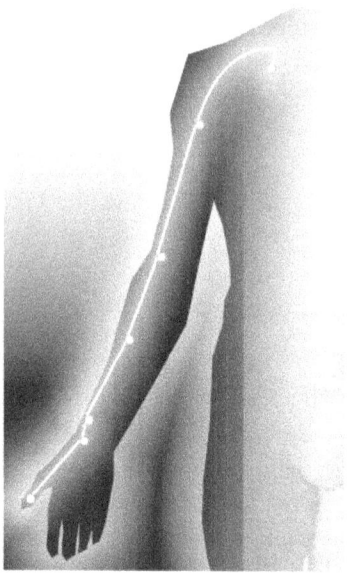

Fig. 27. Meridian of the lungs

Critical periods for circulation of the energy along the meridian in the year are the spring months. Then the early warming of the weather and spring florescence wake up all lingering traumas from former diseases of the respiratory system and block the normal flow of energy.

Like the meridian of the lungs all remaining meridians have energy points, playing an exceedingly important function. These points are called cardinal or standard points of the meridian. More important standard points of the lung meridian are:

Analeptic - P9
Sedative - P5
Auxiliary - P9

Sympathetic -   V13
Signal -   P1
Pain allaying -   P6
Stabilizing to the meridian of the colon - P7.

Meridian of the colon

Polarity - YAN. The movement of the energy is centripetal from the meridian of the lungs to the meridian of the stomach. It consists of 20 energy points. (Fig. 28)

I organize the remote bioenergy channel and pass upward from the outer side of the tip of the forefinger as I follow the respective twenty points of the meridian – to the shoulder on one side along the neck and upwards to the lower part of the nose.

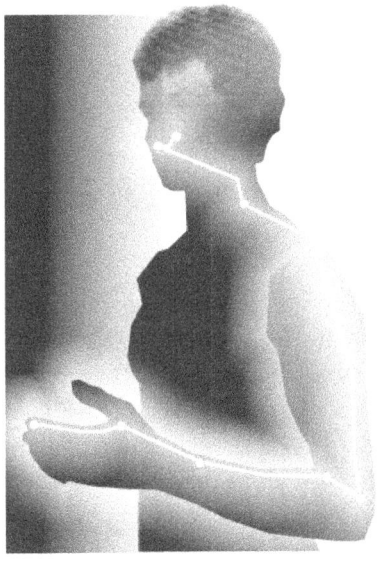

Fig. 28. Meridian of the colon

Disharmony in this meridian cause diseases like: colitis, gastritis, ulcers of the stomach and the duodenum, hemorrhoids, spondylosis of the cervical vertebrae, plexitis, periarthritis, diseases of the gums and teeth, of the throat and larynx, of the antrums.

With normal flow of the energy in the meridian positively influenced are the digestion and the olfaction, the polarity of IN and YAN is balanced and the breathing is normal and quiet.

The most active period of the meridian is from 5 to 7 o'clock. By using this time and the morning hours for diagnostics quick clearing and balancing of the energy flow in the meridian is achieved.

After take of harmful or inappropriate for the organism food in most cases a strong sharp pain is felt in the intestines at 5 o'clock in the morning.

The colors which influence favorably the auto diagnostics and visualization are the white and light yellow golden color.

A critical period for the circulation of the energy along this meridian is the autumn months. The power of the vibration energy of the channel is enhanced which causes cleansing of the digestive system. Standard acupuncture spots:

Analeptic -   G11
Sedative -   G12
Auxiliary -   G14
Sympathetic -   V25
Signal -   E25
Pain allaying -   G17

Stabilizing to the meridian of the lungs - G16.

Meridian of the spleen

Polarity - IN. The energy movement along it is centripetal. It receives energy from the meridian of the stomach and transmits it to the meridian of the heart. It consists of 21 energy points. (Fig. 29)

I concentrate myself in the first point of the meridian - in the middle of the hallux and I organize the remote bioenergy channel. I continue upwards, following all its points - along the internal side of the leg to the loins at the beginning of the thigh, and after that - upwards along the abdomen to the front side of the chest in the region the lung apex. From there I descend downwards to the armpit.

The interruption of the flow of the energy in the meridian is a symptom which directs me to seek diseases like gastritis, ulcers of the stomach and the duodenum, pancreatitis, colitis, cholecystitis, cysts of the ovaries and irregular catamenial cycle in the women, impotency, anemia and general lassitude of the organism. Traumas in the region of the ankles of the legs also have some effect on the normal flow of the energy.

The active period is from 8 to 11 o'clock.

The energy in this meridian is balanced with difficulty and mostly after the diseases has been eradicated. In the cases when there are traumatized limbs this channel remains broken forever and causes swelling and extumescences in the lower part of the legs.

The colors which help the auto diagnostic visualization are the pure orange-yellow and orange-red hues.

The critical periods are connected with the increase of humidity in nature. Then the power of flow of the energy in the meridian decreases and it is jammed even in starting acupuncture points. This causes extumescences in the lower part of the legs, accompanied by symptoms of pains and fatigue.

Standard acupuncture points:
Analeptic -   RP2
Sedative -   RP5
Auxiliary -   RP3
Sympathetic - V20
Signal -   F13
Pain allaying - RP8
Stabilizing point to the meridian of the stomach - RP4.

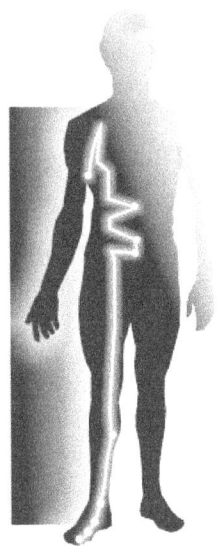

Fig. 29. Meridian of the spleen

Meridian of the stomach

Polarity - YAN. The movement of the energy in it is centripetal and passes through 45 energy points. It receives energy from the meridian of the colon and transmits it to the meridian of the spleen. (Fig. 30)

I organize the remote bioenergy channel by concentrating in the starting acupuncture point, situated in the lower part of the eye. I continue along the acupuncture points along the line of the jaw and around it - downwards along jugulum to the clavicle, cross the chest downwards through the abdomen, inguinal region, I pass through the front part of the leg and I come out of the second toe.

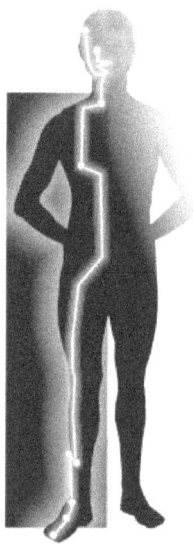

Fig. 30. Meridian of the stomach

The harmonic flow of the energy in the meridian is broken in diseases of the type of facial paralysis, respiratory problems, gastritis, ulcers of the stomach and the duodenum, colitis, cholecistitis and other diseases of the gall bladder and the liver, appendicitis, arthrosis of the thigh and knee joints.

The normal flow of energy in the meridian influences favorably the respiratory and digestive systems, the sight, the gestation sensations, the catamenial cycle in women. It stabilizes the gait.

The most active period is between 7 and 8 o'clock.

The color which influences favorably auto diagnostic visualization is the intense yellow color.

Critical for the meridian are the periods when seasons change. Standard acupuncture spots:

Analeptic -   E41
Sedative -   E45
Auxiliary -   E42
Sympathetic -   V21
Signal -   VC12
Pain allaying -   E34

Stabilizing point in the meridian of the spleen is the acupuncture point - E40.

Meridian of the heart

Polarity - IN. The movement of the energy in it is centrifugal. The energy from the meridian of the spleen passes through 8 acupuncture points of the cardiac meridian, and it transmits it to the meridian of the colon. (Fig. 31)

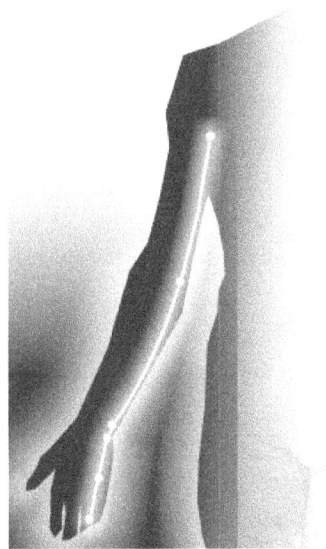

Fig. 31. Meridian of the heart

I concentrate myself in the starting point in the region of the armpit, organize the remote bioenergy channel and I pass with it through the nine acupuncture points of the meridian - from the middle of the internal side of the hand to the little finger.

There is a misbalance in diseases of the respiratory system, the cardio-vascular system, in cases of tendosynovitis, epicondylitis, in states of anxiety, fatigue and insomnia.

The harmonic flow of the energy in the meridian influences favorably the cardiac activity, the blood circulation and perspiration.

The most active period of the meridian is from 11 to 13 o'clock.

In people with problems in the respiratory and cardio-vascular systems this meridian is jammed very easily. In such cases when I harmonize the energy at the beginning of the therapy it is frequently necessary for instance at every ten minutes, to pass again and again, until an uninterrupted and with normal frequency energy flow is created in the meridian.

The colors which influence favorably auto diagnostic visualization are pink and golden.

The most critical period for circulation of the energy in this meridian is in the summer months.

Standard acupuncture points:
Analeptic -   C9
Sedative -   C7
Auxiliary -   C7
Sympathetic -   V15
Signal -   VC14
Pain allaying -   C6
Stabilizing point to the meridian of the intestines - C5.

Meridian of the intestines

Polarity - YAN. It has 18 energy spots. The direction of the energy is centrifugal. It comes from the meridian of the heart and is conveyed to the meridian of the bladder. (Fig. 32)

I concentrate myself in the starting acupuncture point – the tip of the little finger on the outer side of the hand. I organize the remote bioenergy channel and pass ascending upward along the hand. I cross in zigzag the elbow from the side of the back and continue in front at a distance of one inch from the clavicle. Here the information remote channel forms acute angle and rushes upward to the face, reaches on one side from the lower part of the nose and again with acute angle – ends at the cheek bones in front of the ear.

The diseases which cause misbalance of the energy in the meridian are all gastro-intestinal diseases, epicondilitis, bursitis, plexitis, inflammation of the upper respiratory tracts, of the tonsils, of the ears, of the teeth and gums.

The normal flow of energy in the channel influences positively the quick and timely digestion, the auditory and speech activities.

In blocking the energy in this channel strong sharp pains are felt in the intestines in the region of the umbilicus and headache.

The most active period is between 13 and 15 o'clock. This suggests that the most appropriate time for lunch is 13 o'clock.

The colors which influence positively the auto diagnostic visualization are red and navy-blue.

Critical are all cases of food abuse. Eating late in the evening is also dangerous.

Standard acupuncture points:
Analeptic -   IG3
Sedative -   IG8
Auxiliary -   IG4
Sympathetic -   V27
Signal -   VC4
Pain allaying -   IG6
Stabilizing point to the meridian of the heart - IG7.

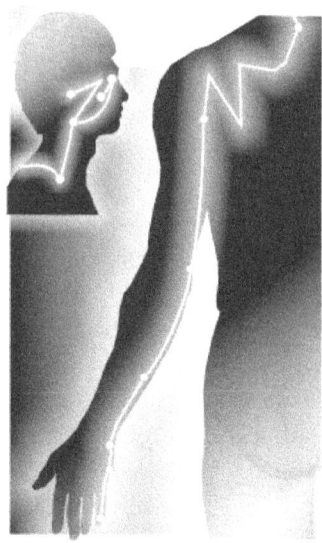

Fig. 32. Meridian of the intestines

Meridian of the kidneys

Polarity - IN. In it the energy moves centripetally, it has 27 energy points. It receives energy from the meridian of the bladder and transmits it to the meridian of the pericardium. (Fig. 33)

I concentrate myself in the starting acupuncture point, which is found in the middle on the inside of the foot. I organize the remote bioenergy channel and through it I climb along the internal side of the leg, I pass through the thigh, the inguinal region, through the abdomen, the stomach, the sternum and pull out the information channel in the region of the clavicle.

A disharmony of the energy in the meridian appears in all diseases of the kidneys, of the genitals, of the intestines, of the pancreas, of the stomach and the gall bladder, of the lungs. Negative influence has also the strong emotions like rage and fear. They block the path of the energy in the meridian from the stomach to its final energy point. With nephric anomalies the energy in the meridian is blocked already in the starting acupuncture points and causes swelling, hardening and cooling of the foot and calf.

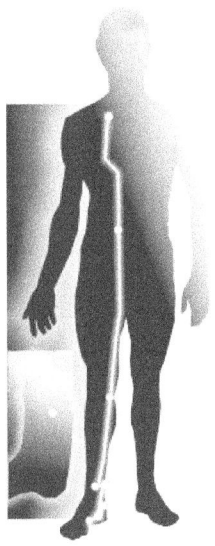

Fig. 33. Meridian of the kidneys

By the normal flow of the energy in the meridian are favorably influenced the diaphragm, the bones, the feet and the whole digestive system.

The most active period for the movement of the energy in the meridian is from 17 to 18 o'clock. In these hours if a man have anomalies in the kidneys he starts to feel some kind of burden and pain there. These hours are favorable for auto diagnostics, auto therapy and phytotherapy.

The colors, which influence favorably the auto diagnostic visualization, are red and intense orange color.

Critical period for the normal circulation of the energy in the meridian are the winter months. Exceedingly important in these periods is to keep the feet warm and dry.

Standard acupuncture points:
Analeptic -   R7
Sedative -   R1
Auxiliary -   R3 (5)
Sympathetic -   V23
Signal -   VB25
Pain allaying -   R5 (4)

Stabilizing point to the meridian of the bladder - R4 (6).

Meridian of the bladder

Polarity - YAN. The movement of the energy is centrifugal and passes through 67 acupuncture points. It comes from the meridian of the intestines and is conveyed to the meridian of the kidneys. (Fig. 34)

In order to make diagnosis of the energy in this meridian, it is best for the patient to lye on his stomach, with the head turned to his left shoulder.

I organize a remote bioenergy channel from the starting acupuncture point - the middle of the eye. I direct it upwards along the head and I follow the sixty seven acupuncture points. They descend vertically through the vertex to the sacrum and up to there are parallel to the managing meridian. They continue downwards along the central part of the thighs, curve slightly in the knee pit and descend to the talus of the ankle and end at the little toe.

A disharmony of the flow of energy in the meridian occurs in all diseases of the excretory system and in these like hemicephalalgia, sinusitis, plexitis, spondiloarthritis, ischioneuralgia and others.

The normal charging of the meridian with energy influences favorably the sight, the audition, the antrums and the bladder.

The active period from 15 to 17 o'clock could be used for auto diagnostic visualization by projecting the passage along the meridian with intense violet-blue color (lavender).

An interruption of the flow of the energy in the meridian is most frequently observed from the ankle to the little toe. This stands exceedingly for women wearing uncomfortable and relatively high shoes, and also for all which stand greater part of the day.

The winter and all cold periods influence unfavorably the energy flow in the meridian.

Standard acupuncture points:
Analeptic -   V67
Sedative -   V65
Auxiliary -   V64
Sympathetic -   V28
Signal -   VC3
Pain allaying - V63
Stabilizing point to the meridian of the kidneys - V58.

The meridian of the bladder it has 12 sympathetic points. These are the points, which are related to the functioning also of other meridians. They are:

V13 - meridian of lungs
V21 - of the stomach
V14 - of the pericardium
V22 - of treble heater
V15 - of the heart
V23 - of the kidneys
V18 - of the liver
V25 - of the colon
V19 - of the gall bladder
V27 - of the intestines
V20 - of the spleen
V28 - of the bladder.

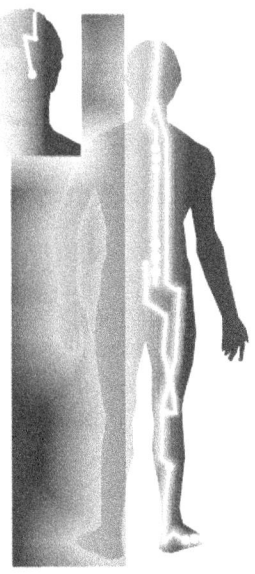

Fig. 34. Meridian of the bladder

Meridian of the pericardium

Polarity - IN. The movement of the energy in it is centrifugal and passes along 8 acupuncture points. It receives energy from the meridian of the kidneys and transmits it to the meridian of the treble heater. (Fig. 35)

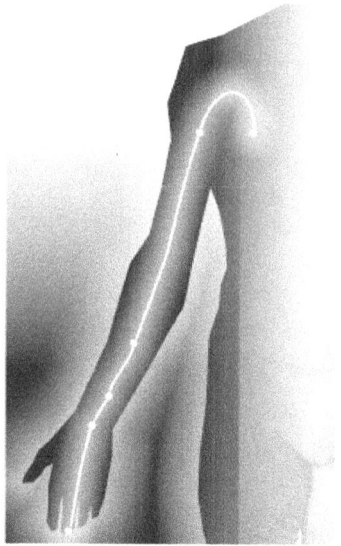

Fig. 35. Meridian of the pericardium

I concentrate myself in the starting point - aside and above the nipple of the breast. I organize the remote bioenergy channel and follow the nine acupuncture points which pass near the armpit and after this along the inner side of the arm I come out to the tip the middle finger.

A misbalance of the flowing energy in the meridian occurs with all diseases in which the cardiac activity is impaired: angina pectoris, heart paralysis, vicium cardi (aortic valve failure), aortic stenosis, mitral stenosis, defects in the cardiac septum, inflammation diseases of the heart, resulting from other contagious diseases and all diseases of the respiratory system – sore throat, cough, tobacco smoking, asthma etc.

The constant circulation of energy in the meridian influences favorably the blood circulation and the breathing. Frequent are the cases when the flow of energy is broken in state of agitation, fear, fright, happiness, tension.

In the cases when for a long time there is no energy flow in the meridian a fillet appears at the base of the palm.

Favorable period of the day for auto diagnostics and auto therapy is from 18 to 21 o'clock.

The position of the meridian gives an opportunity for auto diagnostics by organization of a remote bioenergy channel.

The colors which favorably effect the auto diagnostic visualization are red and green.

Critical for the normal circulation of the energy in the meridian are all very hot and very cold months.

Standard acupuncture points:
Analeptic -   MC9
Sedative -   MC7
Auxiliary -   MC7
Sympathetic -   V14
Signal -   VC17
Pain allaying - MC4

Stabilizing point to the meridian of the treble heater is MC6.

Meridian of the treble heater

Polarity - YAN. The movement of energy in it is centrifugal. It receives energy from the meridian of the pericardium and conveys it to the meridian of the gall bladder. It has 23 active points. (Fig. 36)

I concentrate myself in the tip of the ring finger of the hand and organize the remote bioenergy channel. I pass ascending upward along the posterior outer side of the arm; I pass on the side of the neck, skirt the ears and come out at the end of the brow.

Disharmony occurs with diseases like: plexitis, arthritis, hemicephalgia, spondilitis of the cervical vertebrae, facial paralysis, problems with audition, toothache.

The normal circulation of energy in this meridian helps for sustaining of the body temperature and influences positively the endocrine system and the lymphatic organs.

Active period in the day - between 21 and 23 o'clock.

The colors for auto diagnostic visualization are intense red and violet.

Critical periods in the year are the cold winter months.

Standard acupuncture points:
  Analeptic -   TK3
  Sedative -   TK10
  Auxiliary -   TK4
  Sensitive -   V22
  Signal -   VC5
  Pain allaying - TK7
Stabilizing point to the meridian of the pericardium - TK5
Main points in pathologic problems - VC17, VC12, VC7.

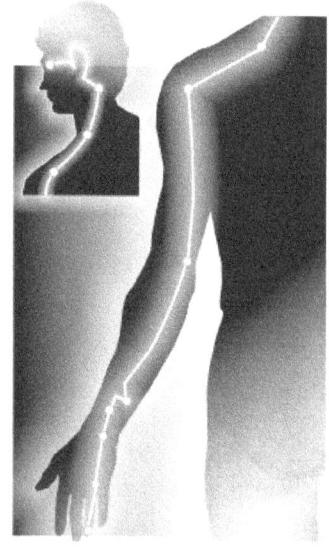

Fig. 36. Meridian of the treble heater

Meridian of the liver

Polarity - IN. The movement of energy in it is centripetal. It receives energy from the meridian of the gall bladder and transmits to the meridian of the lungs. It has fourteen acupuncture points. (Fig. 37)

I concentrate myself in the starting point – the tip of the hallux. I organize the remote bioenergy channel and pass upward along the internal part of the leg. Following the fourteen acupuncture points I reach to the starting point of the meridian of the lungs.

Misbalance in the flow of the energy in the meridian cause all diseases of the liver and most frequently icterus and the jaundice, cholelithiasis, pancreatitis (inflammatory disease of the pancreas), inflammation of the ovaries, different types of colitis, hernia gvinalis etc.

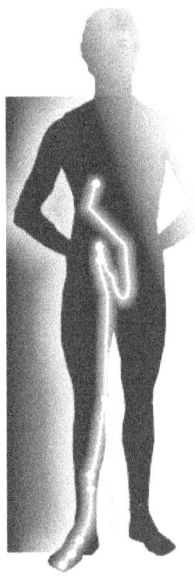

Fig. 37. Meridian of the liver

With normal flow of the energy in the meridian are positively influenced the liver, the muscles, the diaphragm and the digestive system.

Active period of the meridian is from 1 to 3 o'clock.

The color, appropriate for auto diagnostic visualization is turquoise.

Critical for the circulation of the energy in the meridian are the transitional periods, in which weather changes.

Standard points:
Analeptic -   F8
Sedative -   F2
Auxiliary -   F3
Sensitive -   V18
Signal -   F14
Pain allaying - F6

Stabilizing point to the meridian of the gal bladder is F5.

Meridian of the gall bladder

Polarity - YAN. A meridian in which the movement of the energy has centrifugal character. It receives energy from the meridian of the treble heater and transmits it to the meridian of the liver. (Fig. 38)

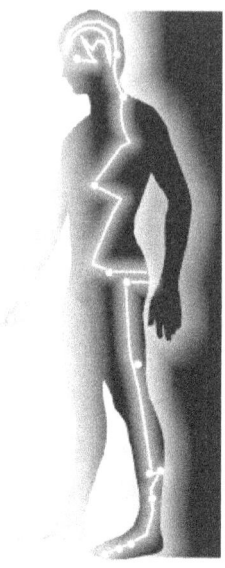

Fig. 38. Meridian of the gall bladder

Through the remote bioenergy channel I follow the forty four points of the gall meridian. It starts aside from the eye, passes through the parietal and temporal bones of the head, descends in a zigzag to the thigh, from there it continues laterally along the ischiatic nerve to the foot and with its last acupuncture point ends in the fourth toe.

A misbalance in the flow of energy in the meridian appears as a result of colds, rhinitis, vestibular syndrome, diseases linked with the sight and audition, head ache, migraines, mostly – in all anomalies and diseases of the gall bladder, the liver and the pancreas.

In normal flow of the energy in the meridian are influenced favorably all organs, which are situated along its pathway.

Most active is the flow of the energy between 11 p.m. and 1 a.m.

Most unfavorable is the cold and windy weather, as well as the cold food.

The colors which influence favorably the auto diagnostic visualization are green (of young spring grass) and turquoise-blue.

Standard acupuncture points:
Analeptic -   VB43
Sedative -   VB38
Auxiliary -   VB40
Sensitive -   V19
Signal -   VB24
Pain allaying -   VB36
Stabilizing point to the meridian of the liver - VB37.

Managing meridian

Polarity - YAN. Controls the energy in all meridians with polarity YAN. The meridian is active in all hours of the day. It has twelve acupuncture points. (Fig. 39)

I organize the remote bioenergy channel by concentrating on the starting point – the extremity apex of the sacrum. I direct it ascending upward along the spinal column. If problems like spondilosis of the vertebrae, displacement of the discs and injury of the intravertebral spaces of the spinal column are absent, strong and constantly pulsating vibrations above the respective twenty seven acupuncture points are felt.

Changes in the flow of the energy in the managing meridian are felt:
- in the presence of diseases of the nervous system, linked with the excretory and genito-urinary system in the points VG1 - VG;
- in the presence of diseases of the nervous system and problems in the gastro-intestinal tract in the points VG5 - VG8;
- in the presence of diseases of the nervous system and the upper respiratory pathways in the points VG8 - VG13;
- in the presence of symptoms of diseases of nervous and the respiratory systems and organs of sight, the olfaction, audition and some psychic diseases in points VG14 - VG28.

The normal flow of energy in the managing meridian has favorable effect of the nervous system in the human organism.

Respectively the misbalance in its energy occurs in the presence of all diseases of the nervous system. From hernia discalis, lumbago, plexitis, neuralgia, paralysis and others - to all psychic diseases, shock and coma.

The position of the meridian gives an opportunity for auto diagnostics through organizing of the remote bioenergy channel. It is well to use the method of visualization. The meridian is filled with clear sky light from the sacrum to the pallet in the buccal cavity.

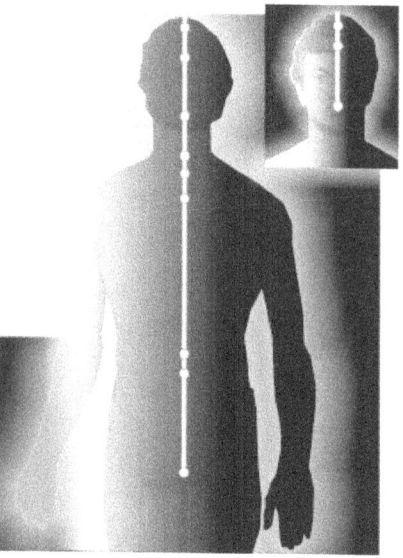

Fig. 39. Managing meridian

Reproductive meridian

Polarity - IN. It unifies all meridians along which flows IN- energy. (Fig. 40)

I organize the remote bioenergy channel by concentrating in the starting acupuncture point – the apex of the sacrum and I direct it upward along the respective twenty four points situated along the middle of frontal part of the body. At the end is the point at the tip of the tongue.

Misbalance in the flow of energy along this meridian is found out in the presence of diseases of the reproductive, the digestive and the respiratory systems. Also in all diseases in the region of the throat and the jugulum: angina, pharingitis, laryngitis, problems of the thyroid gland, logo neurosis etc.

The meridian is the main conductor of energy from the pituitary gland to all glands with internal secretion. It is like generator. If it functions well it feeds with energy the respiratory and cardiovascular systems, controls the male and female hormones, and sustains the body in its biologic optimum.

Visualization of the energy is best to be carried out in a circle: to start from the starting point of the managing meridian, to pass to the final point – the tip of the tongue in the reproductive meridian and to continue downwards along the frontal side to the extremity apex of the sacrum.

All colors of the rainbow influence favorably the performance of auto diagnostics, auto therapy and visualization.

Fig. 40. Reproductive meridian

CHAPTER EIGHT
Diagnostics Of The Chakras

The diagnostics of the chakras is auxiliary method, which gives me an opportunity to confirm the disease of some of the organs and systems and to determine approximately the time this process had started.
Chakra in Sanskrit means "wheel of light" which revolves in our energy system.
"The chakras or the power centers are points of connection in which the energy flows from one energy body of man in other...all these wheels constantly revolve and in the hub or in the open mouth of each one of them constantly is infused power from the superior world. (Leadbeater, C.W. The Theosophical Publishing House, Wheaton, !!: 1827, 4/5.)
In my practical activity like many other healers I feel constantly the specific energy of the seven energy centers and their link with the physical, emotional, mental and spiritual aspect of our being.
The blocking of the energy of the chakras at body level influences the activity of the endocrine and the nervous systems and vice versa.
All diseases of the nervous system are linked also with breaking of the normal flow of energy in the chakras. In inborn diseases like schizophrenia, child cerebral paralysis, epilepsy the constant turn of the energy is permanently broken and it is almost impossible to be recuperated. Only

in rare cases of acquired epilepsy and logo neurosis this could be realized through special diagnostic and therapeutic techniques.

In brain diseases like oligophrenia only the first three chakras interchange energy and with enhanced activity. This explains the specific for this disease mental state, especially during the puberty.

The mental affections on the other side disturb the exchange of energy of the fifth, sixth and seventh chakras.

The endocrine diseases also modify the exchange of energy in the chakras. For example in hyper function of the thyroid gland the energy in the fifth chakra has vibrations with strong repelling effect reaching up to one meter. The diseases of the pancreas and the suprarenal glands block the turn of the energy of the second and third chakras etc.

Every surgical intervention, no matter how successful it is has negative effect on the flow of energy in the chakras. This is applicable most of all for the cases when with the surgical incisions the integrity of the skin or organs has been impaired in an inappropriate place, especially in the abdominal region. After such operations many people suffer from disturbances of the energy in the third chakra. Quite often they tell me: "Here I have something like a small valve, when it is opened, I am well, when it is closed – I've got no vigor".

If the surgeons know and realize the significance of the chakras as basic energy centers they could comply with them also during the respective surgical interventions not to break the turn of energy in the organism of their patients. Sometimes it is sufficient the incision to be half a centimeter upward or downward. At a medical congress in India two surgeons from Bombay read a report on this topic. They shared their experience and observations on the quick recovery of their patients after surgical intervention in the abdominal cavity, when the pathways of the energies flowing in the chakras have not been affected. Doctor Armindo, a gynecologist from Portugal has developed and practices a method for painless birth using his knowledge about the movement of the energy in the chakras.

Through organization and direction of the remote bioenergy information channel - beam to every chakra separately information is extracted about its specific energy. This energy beam plays the role of an original key to studying and balancing the energy of the chakras.

The chakras represent a special energy system. The necessary information about it could be received if one is acquainted and uses the peculiarities of the movement and turn of the energies in the whole body.

Of exceptional significance for the precise diagnostics is to comply with the change of polarity and direction of the energies of men and women. The positive energy is cosmic energy, while the negative we derive from Earth. The direction of the positive energy is from the fontanel to the apex of the sacrum. It must be known that women in daylight, from dawn till sunset, have energy with negative charge. This means that the diagnostics of the chakras in these hours must start from the first chakra and continue upward to the seventh chakra at the fontanel of the head. After sunset, in women the energy of the chakras flows reversely - from the fontanel to the apex of the sacrum and has positive charge. This means that the bioenergy remote channel-beam must follow this energy pathway.

Men in daylight, from dawn till sunset, have energy with positive charge. The diagnostic bioenergy channel in this case will have to be organized from the seventh chakra (the fontanel) and continue downward along the remaining centers to the sacrum. After sunset men change the polarity of their energy and if diagnostics and therapy are to be performed during this time, then the movement of the bioenergy channel along the chakras is from the apex of the sacrum to the fontanel.

In order to become convinced in the change of energy charges in man and woman, you can do something very usual. Take in your right hand a pendulum and place it at about three centimeters above the hands of several men and women separately during the day and after that during the night. You will see that in men during the day it spins clockwise, and in women - vice versa.

Hence follows that it is not advisable to do this diagnostics in the hours about sunrise and sunset when the energy flow of the chakras turns. Then people are feeling slightly tired, feel like sitting for a while, taking deeper breaths, just relaxing. In these hours one shouldn't appoint important meetings and discussions. However these are the most suitable hours for meditation and relaxation.

The daily turn of the positive and the negative charges in the human body is felt more distinctly during the change of seasons. The well known vernal fatigue could also be explained with the energy: the hours of the day increase and the change of charge of the energies is slowed down - the biologic clock must be reset to other energy regime. That is why one feels lassitude, more hours are necessary for rest.

In my practical activity I gradually found out that the pathway of energy in the separate chakras is made difficult mostly in compassionate individuals, in people who suffer very deeply the problems of the others and can even identify themselves with those who suffer. Such syndrome is noticed in mothers when their children are ill: they receive blocking of the energy centers in dependence on the problem of the child. If it is in the respiratory system, the energy of the cardiac chakra and this of the solar plexus is blocked in the mother. I have had similar cases with my patients.

The long-time blocking of the energy in the chakras has consequence linked with the functioning of the organs and the systems in the respective region. Then appears a swelling in the physical body at the place of the chakra which does not function properly.

The diagnostics and the therapy of the chakras through the remote bioenergy channel-beam give us opportunity to eliminate the pains from which people have been suffering for a long time.

During the diagnostics it is important to enfold simultaneously both sides of the chakras. I approach in the following manner:

The patient is standing upright.

I stand on his left side at a distance of up to 50 centimeters from him.

With my both hands I organize two remote bioenergy channels. My right hand directs the channel-beam to the chakras from the back and the left, respectively to the right - remotely monitors the chakras in the frontal side.

In normal flow of the energy in the chakras the information bioenergy channel receives their vibrations. My fingers start to vibrate, to approach or detach them, to rotate in small, hardly detectable spirals or ellipses, according to the place of the energy center, which I monitor.

In diseases of the organs in the region of some of the chakras, the remote bioenergy channels approach my hands from both sides to the diseased place and shorten their length. Also changed is the visual picture, one could distinguish all colors of light with the exception of white, golden and green.

If the chakra is blocked my hands approach to the physical body.

What has to be done in order to open, to unblock the energy in the chakra?

I start to make pressing movements with the tips of my fingers on both sides near to the physical body. I concentrate my whole attention and my look in this energy center. As soon as I feel the organizing of the remote channel-beam I direct it to receive the well known and characteristic for this chakra vibrations. This happens while I make circular helical movements until I am convinced that the energy in the channel increases. After this I proceed in the way as described at the beginning. I create again remote bioenergy channels from the already known distance. I monitor also the visual luminous picture.

As I pass consecutively from the first to the seventh chakra in physically and mentally healthy persons I receive information which there I will try to depict briefly. (Fig. 41)

First chakra

I direct both remote bioenergy channels to the apex of the sacrum.

The first sensation in the tips of my fingers is that the channels-beams join in the specific target. The second sensation is for slight repelling. Along the remote channels energy information starts to flow from the chakra. They are filled with circular motive forces, resembling the passage along winded spring. They detach my fingers to the sides. The detachment depends on the force which man possesses in this energy center. Usually this detachment is from five to ten centimeters.

The visual picture which I can project on the wall with my look is red in different hues. Near the chakra the color is more intense and near the tips of my fingers it changes to pink.

Second chakra

I organize again the two remote channels by concentrating at one distance cun below the umbilicus.

The sensations in the information channel in the first and second moments are the same as depicted herein above during the monitoring of the first chakra.

The diagnostic channel-beam receives information from the energy in the chakra which starts to rotate the fingers of my hands in concentric circles. Their diameter increases to 20 centimeters.

The visual luminous picture, which I project on the wall, is with orange ethereal color.

Third chakra

I concentrate myself in the apex of the sternum. This place in the Bulgarian folk medicine is known with the phrase "in the pit of the stomach".

The information channels join inside and nearly two centimeters downwards. In the point of joining they increase their energy in volume. I receive the vibrations from this chakra with my fingers and my palms. I feel them like a bundle of power lines -beams, which come out from the center of the chakra in both directions.

The visual luminous picture projected on the wall is golden yellow.

Fourth chakra

I concentrate myself in the center of the thymus gland.

The information channels are filled with medium- frequency energy which makes my fingers move in zigzag.

The visual luminous picture is light green.

Fifth chakra

I concentrate myself in the base of the thyroid gland.

After the sensation in the moment of joining both information channels in my fingers receive repelling, but slower pulsations of equal frequency.

The visual luminous picture is intense petroleum blue. The pulsations are luminously reflected as a dotted line. Every trace of it has a length of about 5 millimeters.

Sixth chakra

I concentrate myself in the middle of the forehead and at the same level - in the parietal part of the head. I organize the two information channels. In the point of their joining I feel thermal repulsion. The information channels are filled with energy with repelling pulsations above middle frequency similar to these of the heart. They have radial structure and gradually dissociate the fingers of my hands.

The visual luminous picture is a bright pulsating spot with orange sparkling hue. The pulsations are luminously reflected like a dotted line. Every dot is with length of about 2 millimeters. The more distant it is from the chakra the more pink violet it becomes.

Seventh chakra

I concentrate myself in the fontanel. I organize a remote bioenergy channel with the right hand at a distance to 50 centimeters above it. The bioenergy channel takes the shape of a funnel filled with a bundle of beams with equal frequency of the oscillations. The length the beams in this funnel depends on the time of examination, the sex of the patient and his bodily and mental state in the moment.

The visual luminous picture represents the outstanding contour of the funnel with sparkling orange light. If you look at it more closely you can see that it consists of three luminous layers - white, orange and golden, situated from inside out.

Fig. 41. The chakras or the power centers are points of connection in which the energy flows from one energy body of man in other...all these wheels constantly revolve and in the hub or in the open mouth of each one of them constantly is infused power from the superior world

In order to receive fuller picture and to feel the pouring of the energies from one chakra into the other I CARRY OUT THE FOLLOWING DIAGNOSTIC PROCEDURE.

The patient is lying on his back on a medical couch.

I concentrate myself and with my left hand organize the remote bioenergy channel, by receiving the vibrations of the flowing out energy from the seventh chakra. With my right hand I receive the vibrations from the first chakra. After this the diagnostic channel, organized with this hand, starts upward and passes consecutively through the remaining energy centers. My aim is to equalize the length of the channel and the frequency of the vibrations in all chakras. In this manner is achieved balancing and equalizing of the energy status of the patient.

I want to pay attention to some special details of the information received upon the practicing of this method as with women and men the diagnostic bioenergy channel intercepts different signals.

For example when monitoring the first chakra in the region of the sacrum of a woman, the bioenergy remote channel, apart from the well known characteristics has also vibrations with attracting effect. In men it happens vice versa - the vibrations are with repelling effect.

I find the explanation for this in the structure of the feminine genital system; it is inside in the pelvis. The power of the feminine energy is in the chakras of the solar plexus and this of the thymus gland (known also as cardiac chakra). In these two chakras in women the diagnostic bioenergy channel is filled with vibrations with repelling charge, which increases its length and widen their bioenergy field there. Between these two chakras are found the breasts, they protrude and play the role of vibrating antennae, which balance the energy.

The concentration above the cardiac chakra in men is connected with the reception of attracting vibrations and this shortens the length of the bioenergy channel. With them this chakra is always ready to receive more energy. Observe little boys - they constantly need their mothers. The energy of men overflows in the first chakra and increases the volume of the biofield there. This coincides also with the structure of the male genitals which are outside the pelvis. The vibrations there are above medium frequency, the energy in the remote bioenergy channel increases and could go out at a distance of two meters from the examined patient.

The execution of the depicted diagnostic procedures helps me in the balancing and harmonizing of the energy in the chakras, increases and stabilizes the biofield of the person.

CHAPTER NINE

Use Of The Visualisation In Diagnostics

My activity as a bioenergytherapeutist is impossible without the visualization. In order to make diagnostics of each system, I first concentrate, I visualize it and after this I organize the remote bioenergy channel-beam. The same preparation I make also when the energy meridians from the Chinese energy system are used. I project every energy meridian and its acupuncture points on the body of the examined patient and then I pass along them.

The ability to concentrate myself and visualize the organs and the systems in the organism of man, when I work gives me the opportunity to receive instant color pictures of them. The long practice in deciphering, for reading of the coded in these pictures information, gives me the opportunity to enrich my sensorial perceptibility and to use it for more profound research.

There are cases, when people come to me whom I have treated five or more years ago. Already in the first moment of concentration for organization of the remote bioenergy channel information is sent back to me with all details at psychic and bodily level related to this man. After so many diagnostic examinations I can say that every man has his specific vibration level. It is imprinted in my memory and appears immediately when he again stands in front of me. So inevitably I match my actual observations with the information saved in my memory.

I teach almost all patients to use the visualization for diagnostics. At the beginning we do exercises which I guide, after this everyone could repeat them by himself under certain conditions.

Through visualization the healer could every day diagnose and control the treatment of his patient from a distance. On the fore ground comes the opportunity - in case of complex or sudden diseases - the healer and the patient to program and perform in concord voice the same curative program at different places in precisely determined time of the day.

The practice demonstrates that everyone who knows anatomy and the pathways of every system, through the visualization and with goodwill could spend some time for his health. The visualization helps not only in times of crisis, it could be and it is good to be used also for prophylactics, for stimulation of every system in the human organism.

In my work with athletes I perfected one method for diagnostic visualization which I called "DYNAMIC VISUALISATION". I tested it for the first time on the national team in weight lifting. The coach of the team at that time Nourair Nourikian and the doctor of the team Dr. Petkov took part in it. At the time of carrying out this method, apart from the specific information for every one of the athletes the mental and physical capacities of the team as a whole revealed before us. The boys were educated, intelligent and the visualization helped every one of them to discover his potential and to come to believe in it, and the coaches: to be aware of the limits of the capacities of the contestants. I learned to creates and perfect my program for protection and concentration through the visualization.

The application of the "dynamic visualization" gives me the opportunity to precisely determine the physical state of the athlete and his specific capacities for loading the different parts of the body depending on the type of sport which he practices. I register also his actual psychic characteristic and his capacities for concentration in the very moment. Along with this I manage to feel and analyze his physical and mental state to the moment of examination and to foresee his progress as an athlete.

I shared my results from the dynamic visualization with the athletes in general at the world congress on the "Sylva method" in Athens in 1882. In a conversation with Jose Sylva, he encouraged me by saying: "This method is yours. In order to apply it you use not only the knowledge, which you constantly acquire and recreate, but also your strong intuition and devotion in service of the people".

The perfection and the use of the concentration and visualization became a necessity for me; it turned to be my normal existence. I constantly train my mental activity and clear it from negative thoughts. Every day I try to visualize all positive thoughts and situations connected with the health of the people around me. Sometimes it happens so, that my sensorium gives me ready answers in extreme situations. In 1882 one unpleasant incident happened with the national weightlifting team. Then several boys were accused that they had taken prohibited stimulating preparations and could not participate in the forthcoming international contest. I heard the news on the radio when I was going to work, and it hurt me because I felt with my whole being that this was a provocation. After an hour I had already forgotten I was treating another man also an athlete, who asked me whether I had heard the news. Without asking what news he is talking about in front of my eyes appeared the images of three of the accused contestants to drink diluted yoghurt in front of one big refrigerator. Immediately I contacted the coach and told him about my vision. From him I understood that exactly these contestants whom I have visualized to drink diluted yoghurt gave positive results research in the tests for doping. The inquest of their cases proved that they are not to blame, it was really was a provocation.

The process of visualization of every anatomic system during the diagnostics charges it with energy. The intensity of the energy in the electric body is increased and then it can become visible even with naked eye. Very frequently I do this in order to show it to my patients. The visual color images which I receive and can project on the white wall during diagnostics of the separate systems are result of the concentration and the visualization. After the energy increase in this manner in the respective anatomic system, it receives it and starts to emit its vibrations specific for the moment, for the person, for the situation. So after long exercises every trained man will be able to see pictures, similar to color scanner in front of him and will be able to project them with his look.

In practice it comes out that the biofield could be increased through the visualization as well. Everyone who decides to perfect himself spiritually must realize its creative function.

Exemplary techniques for auto diagnostic visualization

In order to perform the visualization a special preparation is necessary. In the first place – one must get acquainted with the anatomic position, structure and functions of each system. It is advisable to know the techniques of the Sylva Method and other known methods for relaxation.

Let's try to do diagnostics of the digestive system. On its functioning are dependent also our energy balance and our blood pressure, even our moods.

Be seated comfortably in a chair or lye down with a higher pillow. Switch on your favorite music.

Start inhaling and exhaling and gradually diminish the force and depth of inhaling. After you feel, that you inhale and exhale lightly and hardly noticeably, imagine that soft, intense orange light rushes in your mouth from the outside. You can imagine it like an orange cloud, which penetrates inside in the buccal cavity, fills it, passes downward centimeter by centimeter in the esophagus, envelops it, and penetrates in its structure. It rushes down the stomach, in the pylor, fills gradually the duodenum, the intestines, and the colon. Fills all its folds and comes out of the rectum. If along its pathway downwards the orange light, which you visualize, changes its color in darker hues to brown and black, this means that precisely at these places you have problems. The places, where the orange light passes in color ochre coincide with the accumulation of waste matters. And the places where it turns brown fogging coincide with the problems of the mucous membrane of the gastro-intestinal tract.

After you learn the technique you will be able to repeat it several times a day with curative aim.

In this manner could be used for diagnostic visualization and treatment different lights for the different anatomic systems of the human body. For example for diagnostics of the lungs I use dotted light, although it would be better to make it with bright green light, which corresponds to the energy center of the fourth chakra. However for diagnostics and treatment through visualization I never use green light. It is the light of growth, of germination. And according to me if there is some problem in the system, it could be withheld or developed. For the nervous system I use violet light, for the reproductive system - red light, for the sensory organs - golden light, for the cardio-vascular system - blue light, for the endocrine system - white and amethyst light.

Frequently when we practice the visualization for diagnostic and healing effects, appear lights and light effects which we do not expect. In projecting of orange light for diagnostics of the digestive system in some cases there will be places, where the orange light will pass in white. This takes place when the body is exhausted and tense, and when the stomach is full and there is congested food in the colon.

It is best to do the visualization on an empty stomach and after reduction diet.

I have already written herein above about the significance of the darker hues of the orange light. In visualization with yellow and orange light the places where there are pathologic changes, are colored in dark brown, dark cherry and black. If I use for visualization blue and lilac light along the pathway of the system they could darken to black.

The use of the visualization for diagnostics of the anatomic systems could be combined also with the time of activity of the respective energy meridians in the human body.

CHAPTER TEN

Techniques For Organising And Balancing Of The Energy

I propose two techniques which could easily be mastered and their practicing does not require special conditions and time. On the other hand the benefit from them is great. For many of my patients they have become daily practice.

THE FIRST TECHNIQUE is for quickly organizing and harmonizing of the energy in the chakras, with which is achieved balancing of the movement of the energy flows in the whole body and an opportunity to preserve the energy balance and self-protection.

It is best to practice it in the morning before one gets up from bed.

The essence is in organizing, mobilizing and leveling of the energy in the seven energy centers of the body, so that it will function in synchrony.

This is done as with the help of the thought the attention is focused consecutively on every one of the seven chakras. It is performed with closed eyes; the hands are clasped in fists. In the passage above every chakra its number is pronounced and so thrice from one to seven. In the end of each counting we open our eyes and disengage our palms.

Women start visualization in the first passage along the chakras from below - upward.

First visualization

ONE – coincides with the apex of the sacrum.

TWO - slightly (two centimeters) below the umbilicus.

THREE - the solar plexus.

FOUR – the thymus gland (cardiac chakra).

FIVE - the thyroid gland in the base of the jugulum.

SIX - in the middle of the forehead known as "the third eye"-centers of the pituitary gland.

SEVEN - the fontanel, in the middle of soft part of the head of all new-born.

The second visualization starts from above - downwards again with closed eyes and clasped in fists palms.

The third (last) visualization is repetition of the first.

Men start the first visualization along the respective centers from above - downwards. The second is from below - upwards, and the third is repetition of the first.

The triple repetition of the procedure is not as an end in itself. During the first passage along the chakras is helped and is accompanied the penetration of the energy in the body. In the second - the energy levels and stabilizes, and in the last it is harmonized and the body gets ready for the day.

This technique could be improved and used for diagnostics and prophylactic as when it is carried out the colors, accepted to correspond to the respective chakras, are visualized. They are:

For the first chakra – red

for the second - orange

for the third - golden

for the fourth - green

for the fifth - blue

for the sixth - purple

for the seventh - white

The technique is carried out with closed eyes and inside look, directed in the middle of the forehead between the brows. At this place appears a luminous display with a black speck in the middle, which receives first the basic color of our electric (ethereal) body in the moment. After several minutes of visualization of the red color in the first chakra, the display must fill with this color. This hints that we have succeeded to carry out the process of visualization and in this region of the body there are no anatomic and energy problems.

In the beginning when you try to visualize the colors in the respective chakras you could see tentative colors or their hues. For example during visualization of red color in the first chakra you can see pink or orange. The aim is for you to concentrate your attention there until you manage to see the red color.

If in the course of twenty days you perform the technique and do not succeed to visualize the red color it would be better to call a specialist in order to make tests of your urinary and reproductive systems. You should practice in such a way also with the projection of the remaining colors, specific for chakras.

For the projection of the colors could be chosen the appropriate moment in the day or in the evening, in which man is most calm and is not anxious because of the lack of time.

You can repeat the concentration on chakras several times in standing or seated position, if you feel nervous or if you have a headache.

Technique for opening of the fourteen energy channels

Lie comfortably on your back. Balance your breathing for several minutes. Close your eyes and imagine that you concentrate your look in the middle of your forehead. After you exhale the air instantaneously stretch your hands – clasped in fists, your legs and your head. Hold on like this for several seconds. Feel your body stretched to the utmost. After this relax and repeat again the same up to 5 times.

After you carry out this technique the body starts to warm. You receive sensation for the passage of ants on your palms and soles, on the fingers and toes. The energy in the body increases, the electric counterpart shines with bright blue light.

It is advisable this technique to be performed in the morning and before every forthcoming mental and physical load.

This technique helps me personally even when I can't get sleep due to over fatigue. Then the energy in the body cannot circulate normally. After repetition of the exercise my body is released from the tension and starts to relax for sleep.

What are the practical results from the carrying out of the depicted herein above techniques

They harmonize the energy; stabilize the oscillations of the brain waves in the working beta level. Stimulate the activity of the endocrine system because every energy center is in contact also with the respective gland with internal secretion.

They stabilize the biofield and secure the mental self-protection in stress situations. They safeguard from draining of your energy by other persons, from spells and other pernicious effects.

After tenacious twenty day performance of the first technique, the lights in constitutive bodies of the aura clear and it starts to shine. One feels self-confident - fit for work and organized.

This technique prevents children from stress; when they must be waken by compulsion. This could be performed by every mother and child as a game, until the child is fully awake. In Bulgaria 10 years ago I made acquaintance with several children 4-5 years of age with epileptic syndrome. Their first fit coincided with compelled wakening early in the morning. Their parents, seized by the stress not to be late for work, and the child - for the kindergarten, had not taken in mind at all how this would affect a child in deep sleep. The results are convulsion and deviations from the normal limits in the oscillations of the brain waves. In 1881 together with assistant professor Tocheva we monitored the results from performance of the technique by stricken by epilepsy children. After the examinations with electro encephalographer, the results of all and every child had significant improvement and had reached the limits of the normal vibrations.

The technique is the most rapid and best defense for every medic, for every healer. It could be used by every surgeon, before the beginning of surgical intervention and immediately after this. This preventive measure is necessary also for every doctor with the change of patients. It will prevent him from contacting and conveying the diseases from one patient to another. The strong, shining, organized and harmonized biofield of the therapist increases his sensitivity, sharpens his intuition and he manages to notice immediately the essence of the problem, he sees this, which others miss.

In the process of my practice I realized that one of the most nonorganised bio-fields is the one of the teachers. They are exposed to constant attack of the critical looks of their students, which is the easiest way to drain the energy. In their aspiration to keep up the order and to teach the new lesson, they strain their emotional body so much, that after work they feel tired to the utmost. The use of this technique before every class will stabilize the biofield of the teacher to such an extent that when he enters the class room, the students will feel its power. His voice will nail their attention and they unnoticeably, without any resistance, will be inclined to receive the information which is taught.

The performance of the technique before and during examination gives strength and energy for full concentration and creative use of the knowledge.

It could be used with success by players, politicians, pilots and every day by all drivers - before they turn the key of their car.

CHAPTER ELEVEN

Technique For The Projection Of Light And Increasing The Energy Field

We all have admired the sunset at the seaside. Unwillingly we have closed our eyes and have paid no attention that we see several more luminous circles – several more suns with green light, which vibrate and change the green color in different hues. When after this we open our eyes and look at the sea, there we see also circles of light, which we have assumed from the sun. Our eyes have projected them in the sea. The same will happen when we look at the horizon. There we shall see again multiple suns projected by our eyes. If we are in a strong energy place by the sea and observe the sunset only for a few minutes we shall find ourselves among thousands shining energy balls, which have covered the sea, the horizon and we are the centers of this pageant. The same could happen in the mountain.

Once you realize this capacity of yours - to project the light, you will be able to use it everywhere. From the moment of assuming the light from the sunrise or from the sunset, to the closing of the eyes and seeing this light inside us we unconscientiously organize our internal energy. In the next instant - with the opening of the eyes and looking ahead – the projection of this light takes place. Its force has diminished, because we have taken from it and return it to nature. That is why the luminous balls which we project fade away and shine in pink. Whatever we look at with this projected light be it a man, a flower, a tree, we shall help it to increase its energy.

This is one turn, exchange of luminous energy, which you assume from the sun and accumulate it in the nature around you. This luminous energy has curative force, which you can use when you project it on the places where there are pains or old traumas.

I will give you an example from my everyday life. These days I was standing in the waiting room at the surgery of a dentist together with one woman with swollen knees. In front of me was a lit by the sun small garden with grass and orange small flowers – like suns. I fixed this picture through the open door of the waiting room and moved my look to the opposite white wall. I saw a rectangle with pink light and blue flowers. I moved my look to the knees of the woman, together with the pink light from the grass and blue flowers and kept it there until the image of the picture faded away. At this moment the internal door opened and the dentist asked the woman in, it was her turn. Standing up from the chair she exclaimed with surprise: "Oh, how easily I got up this time."

Becoming aware of and mastering the technique for projection of the light we can help ourselves, our kin, the flowers in the garden, to direct our look with kindness and tenderness to nature, which surrounds us.

SCIENTIFIC RESEARCH CENTER OF MEDICAL BIOPHYSICS (SRCMB)
Sofia, Bulgaria

Head of SRCMB Ignat Ignatov, M.Sc.
(Master in Physics)
Holder of World Prizes for Alternative Medicine and Biophysics
Swiss Prize 2003
Vernadski's Prize 2003
Chizhevski's Prize 2005
Holder of a Diploma for Exceptional Merits in the Development of People's
Medicine of Russia – Prof. Galperin, 2002

CERTIFICATE
of

Elena
Todorova
Bakalova

For demonstrated results from the research in the Scientific and Research-Centre of Medical Biophysics – Sofia, (attached to the protocol)

SCIENTIFIC RESEARCH CENTER OF MEDICAL BIOPHYSICS /SRCMB/
Sofia, Bulgaria
Head of SRCMB Ignat Ignatov M.Sc.
Holder of World Prizes for Alternative Medicine and Biophysics
Swiss Prize 2003
Vernadski's Prize 2003
Chizhevski's Prize 2005
Town of Teteven, bl.Obedinenie, entr.V, apt.1 Tel. +359 – 678 – 33 44
Mail address: Sofia 1111, 32 Nikolai Kopernik st., apt.6
Tel. +359 – 2 – 72 99 19
+359 – 889 – 63 74 20
E-mail: mbioph@dir.bg
www.medicalbiophysics.dir.bg

REPORT
of

Elena Todorova Bakalova

Personal No. 4909200930 Sliven, Bulgaria
Athens, Greece

Research is conducted by the authors' team of Ignat Ignatov, Prof. Anton Antonov, D.Ph.Sc., Tatyana Galabova (1998).

1. SPECTRAL ANALYSIS OF INFLUENCED DEIONIZED WATER:
The research is conducted by the DNES method – differential non-equilibrium energy spectrum (Antonov, Galabova, 1993).

a.) ΔE – alteration in the average energy of the intermolecular bonds (10^{-3} eV)

b.) λ – wave ranges in the spheres of maximum influence (10^{-6} m)

c.) $\Delta Eeff = \Delta Egive - \Delta Etake$ (10^{-3} eV)

RESULTS

No. of research	1	1
Date:		22.12.2006
Way of influence:	direct giving ($\Delta Egive$)	direct taking ($\Delta Etake$)
ΔE	-4.2	3.9
λ		8.9 –13.8
$\Delta Eeff$		- 8.1

The result norm is: $\Delta Egive \leq -1.1.10^{-3}$ eV and $\Delta Etake \geq 1.1.10^{-3}$ eV
The excellent result norm is: $\Delta Egive \leq -3.5.10^{-3}$ eV or $\Delta Etake \geq 3.5.10^{-3}$ eV

2. SELECTIVE HIGH FREQUENCY DISCHARGE (SHFD)
– Perfected Kirlian effect with a transparent electrode (Antonov, 1984)
a.) aura type (corona, halo, nebula, mixed)
b.) aura homogeneity – 100% homogenous
c.) average width -12.0 mm
d.) average effective width (ΔR) – 12.0 mm

The result norm is: $\Delta R = 2.6.5mm$, and for an excellent result – $\Delta R = 8.7 mm$

3. VASODILATATIVE EFFECT (Ignatov, 1995)
a.) temperature difference of a skin section at bioinfluence – 1.6 °C
b.) maximum effect – 36.7 °C

The result norm is 0.6 °C, and for an excellent result – 1.3 °C. The accounting is made in relation to 35.1 °C.

FINAL ESTIMATE OF BIOINFLUENCE
of
Elena Todorova Bakalova

	estimate:	weight:
1. Alteration of the energy of the intermolecular connections in water.		
a) Bioinfluence in "give" mode	6.00	0.3
b) Bioinfluence in "take" mode	6.00	0.3
2. SHFD (Perfected Kirlian effect)	6.00	0.2
3. Vasodilatative effect	6.00	0.2

Final estimate: Excellent (6.00)

CONCLUSIONS AND RECOMMENDATIONS: An excellent alternative effect is achieved with the spectral analysis of water. The perfected Kirlian effect is with excellent parameters. The vasodilatative effect also demonstrates excellent results. Elena Basakova possesses excellent bioinformational abilities. They are I class.

Head of SRCMB:
Ignat Ignatov (Master in Physics)

Note: The research method is copyrighted.
No responsibility is undertaken for experiments of Elena Todorova Basakova and the scientific persons related to her, outside the territory of the Scientific and Research Centre of Medical Biophysics.

WORLD FEDERATION OF HEALING

Registered Charity No. 1068734

This is to certify that

Elena Bakalova

is a practising healer

being a member of this Federation is bound to uphold the highest professional standards
and to abide by the Code of Ethics and drawn by the Federation

W.F.H.
REG SEAL
VALID TO
SEPT 2001

This certificate is not valid unless
it carries a current W.F.H. seal.

Chairperson
C. K. Harris

Date
Nov 2000

Membership Secretary,

Reg No
H 2686